Moving Target Defense II

Advances in Information Security

Sushil Jajodia

Consulting Editor
Center for Secure Information Systems
George Mason University
Fairfax, VA 22030-4444
email: jajodia@gmu.edu

The goals of the Springer International Series on ADVANCES IN INFORMATION SECURITY are, one, to establish the state of the art of, and set the course for future research in information security and, two, to serve as a central reference source for advanced and timely topics in information security research and development. The scope of this series includes all aspects of computer and network security and related areas such as fault tolerance and software assurance.

ADVANCES IN INFORMATION SECURITY aims to publish thorough and cohesive overviews of specific topics in information security, as well as works that are larger in scope or that contain more detailed background information than can be accommodated in shorter survey articles. The series also serves as a forum for topics that may not have reached a level of maturity to warrant a comprehensive textbook treatment.

Researchers, as well as developers, are encouraged to contact Professor Sushil Jajodia with ideas for books under this series.

For further volumes:
http://www.springer.com/series/5576

Sushil Jajodia • Anup K. Ghosh • V.S. Subrahmanian
Vipin Swarup • Cliff Wang • X. Sean Wang
Editors

Moving Target Defense II

Application of Game Theory and Adversarial Modeling

 Springer

Editors
Sushil Jajodia
George Mason University
Fairfax, VA, USA

Anup K. Ghosh
George Mason University
Fairfax, VA, USA

V.S. Subrahmanian
University of Maryland
College Park, MD, USA

Vipin Swarup
The Mitre Corporation
McLean, VA, USA

Cliff Wang
U.S. Army Research Office
Triangle Park, NC, USA

X. Sean Wang
Fudan University
Shanghai, China

ISSN 1568-2633
ISBN 978-1-4899-9316-8   ISBN 978-1-4614-5416-8 (eBook)
DOI 10.1007/978-1-4614-5416-8
Springer New York Heidelberg Dordrecht London

Preface

One of the most vexing problems in defending against computer intrusions is the seemingly endless supply of exploitable software bugs that exist despite significant progress in secure software development practices. At least once a month (e.g., on Patch Tuesday), major software vendors publish patches that fix the vulnerabilities in their deployed software code base that have been discovered. These patches are often published after the vulnerabilities are known and have been exploited, in some cases for months and years. In currently deployed systems, the attacker has a static target to study and find vulnerabilities, and then a window of exposure to exploit the vulnerability to gain privileged access on other people's machines and networks, until the exploit is noticed, vulnerability found, patch released, and then applied widely. The dynamics of this process significantly favors the attacker over the defender because the attacker needs to find only a single exploitable bug while the defender must ensure none exist. The attacker has plenty of time to analyze the software code, while the defender does not know when the attacker will strike. And finally, the defender typically can only block the exploit once the exploit or vulnerability is known, giving the attacker an automatic advantage in gaining access with zero-day vulnerabilities.

Against this backdrop, the topic of moving target defenses (MTDs) was developed to level the playing field for defenders versus attackers. The basic concept of MTD is to dynamically vary the attack surface of the system being defended, thus taking away the adversary's advantage of being able to study the target system offline and find vulnerabilities that can be exploited at attack time. MTD systems offer probabilistic protections despite exposed vulnerabilities, as long as the vulnerabilities are not predictable by the adversary at the time of attack. MTD has been identified as one of the four key areas of thrust in the White Houses strategic plan for cyber security research and development.

In the first volume of MTD, we presented papers on MTD foundations, MTD approaches based on software transformations and network and software stack configurations. In this follow-on second volume of MTD, a group of leading researchers describe game–theoretic, cyber maneuver, and software transformation approaches for constructing and analyzing MTD systems.

Acknowledgments

We are extremely grateful to the numerous contributors to this book. In particular, it is a pleasure to acknowledge the authors for their contributions. Special thanks go to Susan Lagerstrom-Fife, senior publishing editor for Springer, and Courtney Clark, editorial assistant at Springer for their support of this project. We also wish to thank the Army Research Office for their financial support under the grant number W911NF-10-1-0470.

Fairfax, VA

Sushil Jajodia
Anup K. Ghosh
V. S. Subrahmanian
Vipin Swarup
Cliff Wang
X. Sean Wang

About the Book

The chapters in this book present a range of MTD challenges and promising solution paths based on game–theoretic approaches, network-based cyber maneuver, and software transformations.

In Chap. 1, Manadhata explores the use of attack surface shifting in the moving target defense approach. The chapter formalizes the notion of shifting a software systems attack surface, introduces a method to quantify the shift, and presents a game–theoretic approach to determine an optimal moving target defense strategy. In Chap. 2, Jain et al. describe the challenging real-world problem of applying game–theory for security and present key ideas and algorithms for solving and understanding the characteristics of large-scale real-world security games, some key open research challenges in this area, and exemplars of initial successes of deployed systems. In Chap. 3, Bilar et al. present a detailed study of the coevolution of the Conficker worm and associated defenses against it and a quantitative model for explaining the coevolution. This study demonstrates, in a concrete manner, that attackers and defenders present moving targets to each other since advances by one side are countered by the other. In Chap. 4, Gonzalez summarizes the current state of computational models of human behavior at the individual level, and it describes the challenges and potentials for extending them to address predictions in 2-player (i.e., defender and attacker) noncooperative dynamic cyber security situations.

The next two chapters explore cyber maneuver in network contexts. In Chap. 5, Torrieri et al. identify the research issues and challenges from jamming and other attacks by external sources and insiders. They propose a general framework based on the notion of maneuver keys as spread-spectrum keys; these supplement higher-level network cryptographic keys and provide the means to resist and respond to external and insider attacks. In Chap. 6, Yackoski et al. describe an IPv6-based network architecture that incorporates cryptographically strong dynamics to limit an attacker's ability to plan, spread, and communicate within the network.

The remaining chapters present MTD approaches based on software transformations. In Chap. 7, Le Goues et al. describe the Helix Metamorphic Shield that continuously shifts a program's attack surface in both the spatial and temporal dimensions and reduces the program's attack surface by applying novel evolutionary algorithms to automatically repair vulnerabilities. The interplay between shifting the attack surface and reducing it results in the automated evolution of new program variants whose quality improves over time. In Chap. 8, Jackson et al. review their automated compiler-based code diversification technique, present an in-depth performance analysis of the technique, and demonstrate its real-world applicability by diversifying a full system stack. Finally, in Chap. 9, Pappas et al. describe in-place code randomization, a software diversification technique that can be applied directly on third-party software. They demonstrate how in-place code randomization can harden inherently vulnerable Windows 7 applications and provide probabilistic protection against return-oriented programming (ROP) attacks.

Contents

Chapter 1
Game Theoretic Approaches to Attack Surface Shifting

Pratyusa K. Manadhata

Abstract A software system's attack surface is the set of ways in which the system can be attacked. In our prior work, we introduced an attack surface measurement and reduction method to mitigate a software system's security risk (Manadhata, An attack surface metric, Ph.D. thesis, Carnegie Mellon University, 2008; Manadhata and Wing, IEEE Trans. Softw. Eng. 37:371–386, 2011). In this paper, we explore the use of *attack surface shifting* in the *moving target defense* approach. We formalize the notion of shifting the attack surface and introduce a method to quantify the shift. We cast the moving target defense approach as a security-usability trade-off and introduce a two-player stochastic game model to determine an optimal moving target defense strategy. A system's defender can use our game theoretic approach to optimally shift and reduce the system's attack surface.

1.1 Introduction

In our prior work, we formalized the notion of a software system's *attack surface* and proposed to use a system's attack surface measurement as an indicator of the system's security [5, 6]. Intuitively, a system's attack surface is the set of ways in which an adversary can enter the system and potentially cause damage. Hence the larger the attack surface, the more insecure the system; we can mitigate a system's security risk by reducing the system's attack surface. We also introduced an *attack surface metric* to measure a system's attack surface in a systematic manner.

Our prior work focused on the uses of attack surface measurements in the software development process. We introduced an *attack surface reduction* approach that complements the software industry's traditional code quality improvement approach to mitigate security risk. The code quality improvement effort aims toward

P.K. Manadhata (✉)
HP Labs, #301, 5 Vaughn Dr, Princeton, NJ 08854, USA
e-mail: manadhata@cmu.edu

S. Jajodia et al. (eds.), *Moving Target Defense II: Application of Game Theory and Adversarial Modeling*, Advances in Information Security 100, DOI 10.1007/978-1-4614-5416-8_1, © Springer Science+Business Media New York 2013

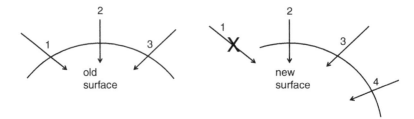

Fig. 1.1 If we shift a system's attack surface, then attacks that worked in the past, e.g., attack 1, may not work any more. The shifting, however, may enable new attacks, e.g. attack 4, on the system

reducing the number of security vulnerabilities in software. In practice, however, building large and complex software devoid of security vulnerabilities remains a very difficult task. Software vendors have to embrace the hard fact that their software will ship with both known and future vulnerabilities in them and many of those vulnerabilities will be discovered and exploited. They can, however, minimize the risk associated with the exploitation of these vulnerabilities by reducing their software's attack surfaces. A smaller attack surface makes the vulnerabilities' exploitation harder and lowers the damage of exploitation, and hence mitigates the security risk.

In this paper, we focus on the uses of attack surface measurements in the context of *moving target defense*. We consider a scenario where system defenders, e.g., system administrators, are continuously trying to protect their systems from attackers. Moving target defense is a novel protection mechanism where the defenders continuously *shift* their systems' attack surfaces to increase the attacker's effort in exploiting their systems' vulnerabilities [1]. As shown in Fig. 1.1, if a defender shifts a system's attack surface, then old attacks that worked in the past, e.g., attack 1, may not work any more. Hence the attacker has to spent more effort to make past attacks work or find new attacks, e.g., attack 4. We view the interaction between a defender and an attacker as a two-player game and hence explore the use of game theory in shifting the attack surface.

The rest of the paper is organized as follows. We briefly discuss our attack surface measurement approach in Sect. 1.2. In Sect. 1.3, we formalize the notion of shifting the attack surface and discuss the uses of attack surface shifting in moving target defense. In Sect. 1.4, we explore game theoretic approaches to attack surface shifting to achieve an optimal balance between security and usability. We conclude with a summary in Sect. 1.5.

1.2   Attack Surface Measurement

We know from the past that many attacks, e.g., exploiting a buffer overflow, on a system take place by sending data from the system's operating environment into the system. Similarly, many other attacks, e.g., symlink attacks, on a system take place because the system sends data into its environment. In both these types of attacks, an attacker connects to a system using the system's *channels* (e.g., sockets), invokes the system's *methods* (e.g., API), and sends *data items* (e.g., input strings) into the system or receives data items from the system. An attacker can also send (receive) data indirectly into (from) a system by using shared persistent data items (e.g., files). Hence an attacker uses a system's methods, channels, and data items present in the system's environment to attack the system. We collectively refer to a system's methods, channels, and data items as the system's *resources* and thus define a system's attack surface in terms of the system's resources.

1.2.1 Attack Surface Definition

Not all resources, however, are part of the attack surface. A resource is part of the attack surface if an attacker can use the resource in attacks on the system. We introduced the *entry point and exit point framework* to identify these relevant resources.

1.2.1.1 Entry Points

A system's codebase has a set of methods, e.g., the system's API. A method receives arguments as input and returns results as output. A system's methods that receive data items from the system's environment are the system's entry points. For example, a method that receives input from a user or a method that reads a configuration file is an entry point. A method *m* of a system *s* is a *direct entry point* if either (a) a user or a system in *s*'s environment invokes *m* and passes data items as input to *m*, or (b) *m* reads from a persistent data item, or (c) *m* invokes the API of a system in *s*'s environment and receives data items as the result returned. An *indirect entry point* is a method that receives data items from a direct entry point.

1.2.1.2 Exit Points

A system's methods that send data items to the system's environment are the system's exit points. For example, a method that writes to a log file is an exit point. A method *m* of a system *s* is a *direct exit point* if either (a) a user or a system in *s*'s environment invokes *m* and receives data items as results returned from *m*, or (b) *m* writes to a persistent data item, or (c) *m* invokes the API of a system in *s*'s environment and passes data items as input to the API. An *indirect exit point* is a method that sends data to a direct exit point.

1.2.1.3   Channels

Each system also has a set of channels; the channels are the means by which users or other systems in the environment communicate with the system, e.g., TCP/UDP sockets, RPC end points, and named pipes. An attacker uses a system's channels to connect to the system and invoke the system's methods. Hence the channels act as another basis for attacks on the system.

1.2.1.4   Untrusted Data Items

An attacker uses persistent data items either to send data indirectly into the system or to receive data indirectly from the system. Examples of persistent data items are files, cookies, database records, and registry entries. A system might read from a file after an attacker writes into the file. Similarly, the attacker might read from a file after the system writes into the file. Hence the persistent data items act as another basis for attacks on a system.

1.2.1.5   Attack Surface Definition

A system's attack surface is the subset of the system's resources that an attacker can use to attack the system. By definition, an attacker can use the set, M, of entry points and exit points, the set, C, of channels, and the set, I, of untrusted data items to send (receive) data into (from) the system to attack the system. Hence M, C, and I are the relevant subset of resources that are part of the attack surface and given a system, s, and its environment, we define s's attack surface as the triple, $\langle M,C,I \rangle$.

1.2.2   *Attack Surface Measurement Method*

A naive way of measuring a system's attack surface is to count the number of resources that contribute to the attack surface. This naive method that gives equal weight to all resources is misleading since all resources are not equally likely to be used by an attacker. We estimate a resource's contribution to a system's attack surface as a *damage potential-effort ratio* where *damage potential* is the level of harm the attacker can cause to the system in using the resource in an attack and *effort* is the amount of work done by the attacker to acquire the necessary access rights to be able to use the resource in an attack.

In practice, we estimate a resource's damage potential and effort in terms of the resource's attributes. For example, we estimate a method's damage potential in terms of the method's *privilege*. An attacker gains the same privilege as a method by using a method in an attack, e.g., the attacker gains root privilege by exploiting a buffer overflow in a method running as root. The attacker can cause damage to the system after gaining `root` privilege. The attacker uses a system's channels to connect to a system and send (receive) data to (from) a system. A channel's *protocol* imposes restrictions on the data exchange allowed using the channel, e.g., a `TCP socket` allows raw bytes to be exchanged whereas an `RPC endpoint` does not. Hence we estimate a channel's damage potential in terms of the channel's protocol. The attacker uses persistent data items to send (receive) data indirectly into (from) a system. A persistent data item's *type* imposes restrictions on the data exchange, e.g., a `file` can contain executable code whereas a `registry entry` can not. The attacker can send executable code into the system by using a `file` in an attack, but the attacker can not do the same using a `registry entry`. Hence we estimate a data item's damage potential in terms of the data item's type. The attacker can use a resource in an attack if the attacker has the required *access rights*. The attacker spends effort to acquire these access rights. Hence for the three kinds of resources, i.e., method, channel, and data, we estimate attacker effort to use a resource in an attack in terms of the resource's access rights.

We assume a function, *der*, that maps a resource to its damage potential-effort ratio. In practice, however, we assign numeric values to a resource's attributes to compute the ratio, e.g., we compute a method's damage potential-effort ratio from the numeric values assigned to the method's privilege and access rights. We impose a total order among the values of the attributes and assign numeric values according to the total order. For example, we assume that an attacker can cause more damage to a system by using a method running with `root` privilege than a method running with `non-root` privilege; hence we assign a higher number to the `root` privilege level than the `non-root` privilege level. The exact choice of numeric values is subjective and depends on a system and its environment.

We quantify a system's attack surface measurement along three dimensions: methods, channels, and data. We estimate the total contribution of the methods, the total contribution of the channels, and the total contribution of the data items to the attack surface. Given the attack surface, $\langle M, C, I \rangle$, of a system, s, s's attack surface measurement is the triple $\langle \sum_{m \in M} der(m), \sum_{c \in C} der(c), \sum_{d \in I} der(d) \rangle$.

1.3   Moving Target Defense

In this section, we discuss the uses of attack surface measurements in moving target defense. Moving target defense is a protection approach where a system's defender continuously *shifts* the system's attack surface. Intuitively, the defender may *modify* the attack surface by changing the resources that are part of the attack surface and/or by modifying the contributions of the resources. Not all modifications, however, shift the attack surface. The defender shifts the attack surface by removing at least one resource from the attack surface and/or by reducing at least one resource's damage potential-effort ratio. Everything else being equal, attacks that worked in the past may not work in the future if the attacks depended on the removed (modified) resource. The shifting process, however, might have enabled new attacks on the system by adding new resources to the attack surface. Hence the attacker has to spend more effort to make past attacks work or to identify new attacks.

1.3.1 Shifting the Attack Surface

We formalize the notion of *shifting the attack surface* in this section and introduce a method to quantify the shift. We introduced an I/O automata model of a system and its environment in our prior work; we use the model to define and quantify the shift in the attack surface.

Consider a set, S, of systems, an attacker, U, and a data store, D. For a system, $s \in S$, we define s's environment, $E_s = \langle U, D, T \rangle$, to be a three-tuple where $T = S \setminus \{s\}$ is the set of systems excluding s. U represents the adversary who attacks the systems in S. The data store D allows data sharing among the systems in S and U.

We model a system and the entities present in its environment as I/O automata [4]. An I/O automaton, $A = \langle sig(A), states(A), start(A), steps(A) \rangle$, is a four tuple consisting of an *action signature*, $sig(A)$, that partitions a set, $acts(A)$, of *actions* into three disjoint sets, $in(A)$, $out(A)$, and $int(A)$, of *input*, *output* and *internal* actions, respectively, a set, $states(A)$, of *states*, a non-empty set, $start(A) \subseteq states(A)$, of *start states*, and a *transition relation*, $steps(A) \subseteq states(A) \times acts(A) \times states(A)$. An *execution* of A is an alternating sequence of actions and states beginning with a start state and a *schedule* of an execution is a subsequence of the execution consisting only of the actions appearing in the execution.

Given a system, s, and its environment, E, s's attack surface is the triple, $\langle M, C, I \rangle$, where M is the set of entry points and exit points, C is the set of channels, and I is the set of untrusted data items of s. We denote the set of resources belonging to s's attack surface as $R_s = M \cup C \cup I$. Also, given two resources, r_1 and r_2, of s, we write $r_1 \succ r_2$ to express that r_1 makes a larger contribution to the attack surface than r_2. If we modify s's attack surface, R_o, to obtain a new attack surface, R_n, then we denote a resource, r's, contributions to R_o as r_o and to R_n as r_n. We define attack surface shifting qualitatively as follows.

Definition 1.1. Given a system, s, its environment, E, s's old attack surface, R_o, and s's new attack surface, R_n, s's attack surface has shifted if there exists at least one resource, r, such that (i) $r \in (R_o \setminus R_n)$ or (ii) $(r \in R_o \cap R_n) \wedge (r_o \succ r_n)$.

If we shift s's attack surface, then attacks that worked on s's old attack surface may not work on s's new attack surface. We model s's interaction with its environment as parallel composition, $s \| E$, in our I/O automata model. Since an attacker attacks a system either by sending data into the system or by receiving data from the system, any schedule of the composition $s \| E$ that contains s's input actions or output actions is a potential attack on s. We denote the set of potential attacks on s as $attacks(s, R)$ where R is s's attack surface. In our I/O automata model, if we shift s's attack surface from R_o to R_n, then with respect to the same attacker

Fig. 1.2 If we shift a system's attack surface from R_o to R_n, then at least one attack that worked on R_o will not work any more on R_n

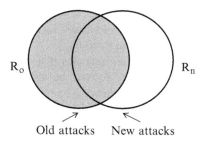

Old attacks   New attacks

and the environment, a few potential attacks on R_o will cease be potential attacks on R_n. Intuitively, if we remove a resource, r, from the attack surface or reduce r's contribution to the attack surface during shifting, then executions of s that contain r will not be executions in the new attack surface. Hence the schedules derived from these executions will not be potential attacks on s in the new attack surface (Fig. 1.2).

Theorem 1.1. *Given a system, s, and its environment, E, if we shift s's attack surface, R_o, to a new attack surface, R_n, then $attacks(s, R_o) \setminus attacks(s, R_n) \neq \emptyset$.*

Proof (Sketch). If we shift s's attack surface from R_o to R_n, then from Definition 1.1, there is at least one resource, r, such that either (i) $r \in (R_o \setminus R_n)$ or (ii) $(r \in R_o \cap R_n) \wedge (r_o \succ r_n)$.

If $r \in (R_o \setminus R_n)$, then without loss of generality, we assume that $R_o = R_n \cup \{r\}$. Since $r \in R_o \wedge r \notin R_n$, following arguments similar to the proof of Theorem 1 in [5], there exists a method, m, such that $m \in R_o \wedge m \notin R_n$. Hence there exists a schedule, β, of the composition $s_{R_o} || E$ containing m such that β is not a schedule of the composition $s_{R_n} || E$. Hence $\beta \in attacks(s, R_o) \wedge \beta \notin attacks(s, R_n)$, and $attacks(s, R_o) \setminus attacks(s, R_n) \neq \emptyset$.

Similarly, if $(r \in R_o \cap R_n) \wedge (r_o \succ r_n)$, then r makes a larger contribution to R_o than R_n. Following arguments similar to the proof of Theorem 3 in [5], there exists a method, $m \in R_o \cap R_n$, such that m has a stronger pre condition and/or a weaker post condition in R_o than R_n. Hence there exists a schedule, β, of the composition $s_{R_o} || E$ containing m such that β is not a schedule of the composition $s_{R_n} || E$. Hence $\beta \in attacks(s, R_o) \wedge \beta \notin attacks(s, R_n)$, and $attacks(s, R_o) \setminus attacks(s, R_n) \neq \emptyset$. $\square$

We introduced a qualitative notion of shifting the attack surface in previous paragraphs. We quantify the shift in the attack surface as follows.

Definition 1.2. Given a system, s, its environment, E, s's old attack surface, R_o, and s's new attack surface, R_n, the shift, ΔAS, in s's attack surface is

$$|R_o \setminus R_n| + |\{r : (r \in R_o \cap R_n) \wedge (r_o \succ r_n)\}|$$

In Definition 1.2, the term $|R_o \setminus R_n|$ represents the number of resources that were part of s's old attack surface, but were removed from s's new attack surface. Similarly, the term

$$|\{r : (r \in R_o \cap R_n) \wedge (r_o \succ r_n)\}|$$

represents the number of resources that make larger contributions to s's new attack surface than the old attack surface. If $\Delta AS > 0$, then we say that s's attack surface has shifted from R_o to R_n.

Our definition assumes that all resources contribute equally to the shift in the attack surface. We may be able to quantify the shift better by considering the resources' attributes, e.g., a resource's damage potential-effort ratio. We leave such quantification approaches as future work.

1.3.2 Ways to Shift the Attack Surface

The defender may modify the attack surface in three different ways. But only two of these three ways shift the attack surface. First, the defender may shift the attack surface and also reduce the attack surface measurement by disabling and/or modifying the system's *features* (Scenario A). Disabling the features reduces the number of entry points, exit points, channels, and data items, and hence reduces the number of resources that are part of the attack surface. Modifying the features reduces the damage potential-effort ratios of the resources that are part of the attack surface, e.g., lowering a method's privilege or increasing the method's access rights, and hence reduces the resources' contributions to the attack surface measurement.

Second, the defender may shift the attack surface by enabling new features and disabling existing features. Disabling the features removes resources from the attack surface and hence shifts the attack surface. The attack surface measurement, however, may decrease (Scenario B), remain the same (Scenario C), or increase (Scenario D). The enabled features increase the attack surface measurement by adding more resources to the attack surface and the disabled features decrease the measurement by removing resources from the attack surface; the overall change in the measurement may be negative, zero, or positive. Similarly, the defender may shift the attack surface by modifying existing features such that the damage potential-effort ratios of a set of resources decrease and the ratios of another set of resources increase. The attack surface measurement may decrease, remain the same, or increase.

Third, the defender may modify the attack surface by enabling new features. The new features add new resources to the attack surface and hence increase the attack surface measurement. The attack surface, however, doesn't shift since the old attack surface still exists and all attacks that worked in the past will still work (Scenario E). The defender may also increase the attack surface measurement without shifting the attack surface by increasing the damage potential-effort ratios of existing resources. We summarize the scenarios in Table 1.1.

From a protection standpoint, the defender's preference over the scenarios is the following: A > B > C > D > E. Scenario A is preferred over scenario B because scenario B adds new resources to the attack surface and the new resources may enable new attacks on the system. Scenario D increases the attack surface measurement; but it may be attractive in moving target defense, especially if the increase in the measurement is low and the shift in the attack surface is large.

Table 1.1 Different scenarios to modify and shift the attack surface. Not all modifications shift the attack surface

Scenario	Features	Attack surface shift	Attack surface measurement
A	Disabled	Yes	Decrease
B	Enabled and disabled	Yes	Decrease
C	Enabled and disabled	Yes	No change
D	Enabled and disabled	Yes	Increase
E	Enabled	No	Increase

1.3.3 A Security-Usability Trade-off

The defender may not always be able to shift and reduce the attack surface. The defender may have to enable new features or modify existing features to provide desirable services to the system's users at the expense of an increased attack surface measurement. For example, the users may demand remote access to the system and hence defender may have to open a new communication channel, e.g., a TCP port, to satisfy the demand. The increased attack surface may enable new attacks on the system. Similarly, attack surface reduction comes at a cost; the reduction process disables or modifies the system's features and hence the system may not be able to provide certain services as before. Hence the defender has to make a classic security-usability trade-off as part of the moving target defense. In Sect. 1.4.1, we discuss game theoretic approaches to determine optimal ways to modify and shift the attack surface to achieve moving target defense.

1.4 Game Theoretic Approaches

In our prior work, we viewed attack surface measurement and reduction as a "static" process, i.e., software developers measure their systems' attack surfaces without making any assumptions about an attacker (e.g., the attacker's resources, skill levels, and behavior) and then try to reduce the attack surface as much as possible. In this section, we consider attack surface reduction and shifting in a dynamic moving target defense environment - a defender continuously tries to protect a system from an attacker by reducing and shifting the attack surface. We model the interaction between the defender and the attacker as a two player game and use game theory to determine *optimal* defense strategies. The game theoretic model allows us to explicitly model the attacker. Hence the defender can choose optimal defense strategies for different attacker profiles such as script kiddies, experienced hackers, organized criminals, and nation states.

1.4.1 Optimal Moving Target Defense

We model the interaction between a system's defender and an attacker as a two-player stochastic extensive game [8].

1.4.1.1 A Game Model

Our stochastic game model is similar to the model introduced by Lye and Wing [3]. Our model is a 7 tuple, $\langle S, A^d, A^a, T, R^d, R^a, \beta \rangle$, where S is a set of system states, A^d is the defender's action set, A^a is the attacker's action set, $T : S \times A^d \times A^a \times S \to [0, 1]$ is the state transition function, $R^d : S \times A^d \times A^a \to \mathbb{R}$ is the defender's reward function where $\mathbb{R}$ is the set of real numbers, $R^a : S \times A^d \times A^a \to \mathbb{R}$ is the attacker's reward function, and $\beta \leq 1$ is a discount factor for discounting future rewards.

The defender and the attacker play the game in the following manner. The system is in the state $s_t \in S$ at time t. The defender performs an action, $a^d \in A^d$, and the attacker performs an action, $a^a \in A^a$. The system then moves to a state, $s_{t+1} \in S$, with probability $T(s_t, a^d, a^a, s_{t+1})$. The defender's reward for performing the action is $r_t^d = R^d(s_t, a^d, a^a)$ and the attacker's reward is $r_t^a = R^a(s_t, a^d, a^a)$. The goals of the defender and the attacker are to maximize their discounted rewards.

1.4.1.2 States, Actions, and Transitions

We model the system as a set, F, of features. F represents the system's features that provide various functionality, e.g., a web server's login feature provides user authentication functionality. A feature can be disabled or if enabled, can be in one of several *configurations*; each configuration is a mapping of state variables to their values. A state, $st \in S$, of the system is a mapping of the features to their configurations, i.e., $st : F \to Configuration$. At a given system state, the defender may choose to shift and reduce the attack surface by *acting* on the features, e.g., the defender may enable a disabled feature, disable an enabled feature, modify an enabled feature's configuration, or leave a feature's configuration unchanged. Hence, a defender action, $a^d \in A^d$, is a mapping of the features to the actions performed by the defender on the features, i.e., $a^d : F \to \{enabled, disabled, modified, unchanged\}$. In a given system state, the defender can choose from a subset of the set of actions A^d. For example, if a feature, f, is in enabled configuration in a state, s, then the defender cannot perform any action that *enables* f; the defender can only disable f, modify f, or leave f unchanged. After the defender performs an action, the system's attack surface changes. The attacker then performs an action to attack the system by utilizing the change in the attack surface; the attacker's action may further enable and disable the system's features. Since the defender's action and the attacker's action enable and/or disable certain system features, the system moves to a new state according to the probabilistic transition function. The specific values of the transition probabilities depend on the system and its operating environment.

Potential state space explosion and potential action space explosion are two disadvantages of our model. The number of states and the number of actions are exponential in the number of features. For simplicity and tractability, we may focus on an important subset of the system's features and may bound the number of features that an action can enable, disable, or modify.

1.4.1.3 Reward Functions

When the defender performs an action, the action may benefit the defender in three ways. First, the defender may provide value to the system's users by enabling features. Second, the defender may mitigate the system's security risk by shifting the attack surface. Third, the defender may mitigate the system's security risk by reducing the attack surface measurement. The action, however, may cost the defender if it disables features or increases the system's attack surface measurement. Hence the defender's reward depends on the change in value derived from the features, the shift in the attack surface, and the change in the attack surface measurement.

Similarly, when the attacker performs an action, the attacker benefits from the increase in the attack surface measurement. But the shift in the attack surface costs the attacker. Hence the attacker's reward depends on the shift in the attack surface and the change in the attack surface measurement.

Consider a state, s, of the system. If the defender performs an action, a^d, in a state, s, and the attacker performs an action, a^a, then we denote the change in the system's features as ΔF, the shift in the attack surface as ΔAS, and the change in the attack surface measurement as ΔASM. Then we model the defender's reward, R^d, and the attacker's reward, R^a, as follows.

$$R^d(s, a^d, a^a) = B_1(\Delta F) + B_2(\Delta AS) - C_1(\Delta ASM)$$
$$R^a(s, a^d, a^a) = B_3(\Delta ASM) - C_2(\Delta AS)$$

B_is and C_is are functions that map the changes in features, attack surface shift, and attack surface measurement to real numbers; the numbers reflect the benefits and costs associated with the changes. The exact choice of B_is and C_is depends on the system and its operating environment. Please note that our choice of reward functions makes the game a general-sum game.

1.4.1.4 Optimal Defense Strategies

We model our game as a complete and perfect information game; each player knows the other player's strategies and pay-offs, and is aware of the game *history*, i.e., the actions already performed by both players in the game [9]. The goal of each player is to maximize their expected discounted pay-off.

A player's *strategy* is a plan of action that the player can take in the game; the strategy specifies the action(s), given the other player's strategy, the player can take at different stages of the game. An optimal strategy maximizes the player's expected pay-off.

A stationary strategy is a strategy that is independent of time and history, and depends only on the system's state. A pure strategy specifies a single action in a state whereas a mixed strategy specifies a probability distribution over possible actions in the state. We use the Nash Equilibrium solution concept to determine the defender's optimal stationary strategy. Filar and Vrieze introduced a non-linear program to find stationary equilibrium strategies in general sum stochastic games [2].

The defender may want an optimal strategy that is dependent on time and history. The strategy specifies optimal defender action(s) after *every* history of the game. Hence the defender can take an optimal action in response to the attacker's action at any time in the game. We use the subgame perfect equilibrium concept to determine the defender's optimal strategy [8]. Murray and Gordon introduced a dynamic programming algorithm to find a subgame perfect equilibrium in general sum stochastic games [7].

Hence the defender can use the concepts of Nash equilibrium and subgame perfect equilibrium to determine optimal strategies for shifting and reducing the attack surface. The optimal strategies enable the defender to make the security-usability trade-off in an informed manner; the system can then provide required services to its users without compromising its security.

1.5 Summary

In summary, we introduced a game theoretic attack surface shifting and reduction approach to achieve moving target defense. System defenders can use our approach to determine their best course of action to protect their systems while providing required services to their systems' users. In the future, we plan to instantiate our model on real world software systems and explore the efficacy of our approach in real world settings.

References

1. National cyber leap year summit 2009 co-chairs report. http://www.cyber.st.dhs.gov/docs/ National_Cyber_Leap_Year_Summit_2009_Co-Chairs_Report.pdf (2009)
2. Filar, J., Vrieze, K.: Competitive Markov decision processes. Springer (1997)
3. Lye, K., Wing, J.M.: Game strategies in network security. International Journal of Information Security pp. 71–86 (2005)
4. Lynch, N., Tuttle, M.: An introduction to input/output automata. CWI-Quarterly **2**(3) (1989)
5. Manadhata, P.K.: An attack surface metric. Ph.D. thesis, Carnegie Mellon University (2008)

6. Manadhata, P.K., Wing, J.M.: An attack surface metric. IEEE Trans. Softw. Eng. **37**, 371–386 (2011). DOI http://dx.doi.org/10.1109/TSE.2010.60. URL http://dx.doi.org/10.1109/ TSE.2010.60
7. Murray, C., Gordon, G.: Finding correlated equilibria in general sum stochastic games. Tech. Rep. CMU-ML-07-113, Carnegie Mellon University (2007)
8. Osborne, M., Rubinstein, A.: A course in game theory. MIT Press (1994)
9. Roy, S., Ellis, C., Shiva, S., Dasgupta, D., Shandilya, V., Wu, Q.: A survey of game theory as applied to network security. Hawaii International Conference on System Sciences **0**, 1–10 (2010). DOI http://doi.ieeecomputersociety.org/10.1109/HICSS.2010.35

Chapter 2
Security Games Applied to Real-World: Research Contributions and Challenges

Manish Jain, Bo An, and Milind Tambe

Abstract The goal of this chapter is to introduce a challenging real-world problem for researchers in multiagent systems and beyond, where our collective efforts may have a significant impact on activities in the real-world. The challenge is in applying game theory for security: our goal is not only to introduce the problem, but also to provide exemplars of initial successes of deployed systems in this problem arena. Furthermore, we present key ideas and algorithms for solving and understanding the characteristics large-scale real-world security games, and then present some key open research challenges in this area.

2.1 Introduction

Security is a critical concern around the world that arises in protecting our ports, airports, transportation or other critical national infrastructure from adversaries, in protecting our wildlife and forests from poachers and smugglers, and in curtailing the illegal flow of weapons, drugs and money; and it arises in problems ranging from physical to cyber-physical systems. In all of these problems, we have limited security resources which prevent full security coverage at all times; instead, limited security resources must be deployed intelligently taking into account differences in priorities of targets requiring security coverage, the responses of the attackers to the security posture and potential uncertainty over the types, capabilities, knowledge and priorities of attackers faced.

Game theory is well-suited to adversarial reasoning for security resource allocation and scheduling problems. Casting the problem as a Bayesian Stackelberg game, new algorithms have been developed for efficiently solving such games that

M. Jain (✉) • B. An • M. Tambe
Computer Science Department, University of Southern California, Los Angeles,
California 90089, USA
e-mail: manish.jain@usc.edu; boa@usc.edu; tambe@usc.edu

S. Jajodia et al. (eds.), *Moving Target Defense II: Application of Game Theory and Adversarial Modeling*, Advances in Information Security 100, DOI 10.1007/978-1-4614-5416-8_2, © Springer Science+Business Media New York 2013

provide randomized patrolling or inspection strategies. These algorithms have led to some initial successes in this challenge problem arena, leading to advances over previous approaches in security scheduling and allocation, e.g., by addressing key weaknesses of predictability of human schedulers. These algorithms are now deployed in multiple applications: ARMOR has been deployed at the Los Angeles International Airport (LAX) since 2007 to randomize checkpoints on the roadways entering the airport and canine patrol routes within the airport terminals [1]; IRIS, a game-theoretic scheduler for randomized deployment of the US Federal Air Marshals (FAMS) requiring significant scale-up in underlying algorithms, has been in use since 2009 [2]; PROTECT, which uses a new set of algorithms based on quantal-response is deployed in the port of Boston for randomizing US coast guard patrolling [3,4]; PROTECT has been deployed in the port of Boston since April 2011 and is now in use at the port of New York; GUARDS is under evaluation for national deployment by the US Transportation Security Administration (TSA) [5], and TRUSTS is being tested by the Los Angeles Sheriffs Department (LASD) in the LA Metro system to schedule randomized patrols for fare inspection [6]. These initial successes point the way to major future applications in a wide range of security arenas; with major research challenges in scaling up our game-theoretic algorithms, to addressing human adversaries' bounded rationality and uncertainties in action execution and observation, as well as in preference elicitation and multiagent learning.

This paper will provide an overview of the models and algorithms, key research challenges and a brief description of our successful deployments. While initial research has made a start, a lot remains to be done; yet these are large-scale interdisciplinary research challenges that call upon multiagent researchers to work with researchers in other disciplines, be "on the ground" with domain experts, and examine real-world constraints and challenges that cannot be abstracted away.

2.2 Stackelberg Security Games

Security problems are increasingly studied using Stackelberg games, since Stackelberg games can appropriately model the strategic interaction between a defender and an attacker. Stackelberg games were first introduced to model leadership and commitment [7], and are now widely used to study security problems ranging from "police and robbers" scenario [8], computer network security [9], missile defense systems [10], and terrorism [11]. Models for arms inspections and border patrolling have also been modeled using inspection games [12], a related family of Stackelberg games.

The wide use of Stackelberg games has inspired theoretic and algorithmic progress leading to the development of fielded applications. These algorithms are central to many fielded applications, as described in Sect. 2.3. For example, DOBSS [13], an algorithm for solving Bayesian Stackelberg games, is central to a fielded application in use at the Los Angeles International Airport [1]. Similarly,

	Defender		Attacker	
Target	Covered	Uncovered	Covered	Uncovered
t_1	10	0	−1	1
t_2	0	−10	−1	1

Table 2.1 Example security game with two targets

Conitzer and Sandholm [14] give complexity results and algorithms for computing optimal commitment strategies in Bayesian Stackelberg games, including both pure and mixed-strategy commitments. This chapter provides details on this use of Stackelberg games for modeling security domains. We first give a generic description of security domains followed by *security games*, the model by which security domains are formulated in the Stackelberg game framework.

2.2.1 Security Domains

In a security domain, a defender must perpetually defend a set of targets using a limited number of resources, whereas the attacker is able to surveil and learn the defender's strategy and attacks after careful planning. This fits precisely into the description of a Stackelberg game if we map the defender to the leader's role and the attacker to the follower's role [12, 15]. An action, or *pure strategy*, for the defender represents deploying a set of resources on patrols or checkpoints, e.g. scheduling checkpoints at the LAX airport or assigning federal air marshals to protect flight tours. The pure strategy for an attacker represents an attack at a target, e.g., a flight. The strategy for the leader is a mixed strategy, a probability distribution over the pure strategies of the defender. Additionally, with each target are also associated a set of payoff values that define the utilities for both the defender and the attacker in case of a successful or a failed attack. These payoffs are represented using the *security game* model, described next.

2.2.2 Security Games

In a security game, a set of four payoffs is associated with each target. These four payoffs are the reward and penalty to both the defender and the attacker in case of a successful or an unsuccessful attack, and are sufficient to define the utilities for both players for all possible outcomes in the security domain. Table 2.1 shows an example security game with two targets, t_1 and t_2. In this example game, if the defender was *covering* (protecting) target t_1 and the attacker attacked t_1, the defender would get 10 units of reward whereas the attacker would receive −1 units.

Security games make the assumption that it is always better for the defender to cover a target as compared to leaving it uncovered, whereas it is always better for the attacker to attack an uncovered target. This assumption is consistent with the

Table 2.2 Example Bayesian security game with two targets and two attacker types

	Attacker Type 1					Attacker Type 2			
	Defender		Attacker			Defender		Attacker	
Target	Cov.	Uncov.	Cov.	Uncov.	Target	Cov.	Uncov.	Cov.	Uncov.
t_1	10	0	−1	1	t_1	5	−4	−2	1
t_2	0	−10	−1	1	t_2	4	−5	−1	2

payoff trends in the real-world. Another crucial feature of the security games is that the payoff of an outcome depends only on the target attacked, and whether or not it is covered by the defender [16]. The payoffs do *not* depend on the remaining aspects of the defender allocation. For example, if an adversary succeeds in attacking target t_1, the penalty for the defender is the same whether the defender was guarding target t_2 or not. Therefore, from a payoff perspective, many resource allocations by the defender are identical. This is exploited during the computation of a defender strategy: only the coverage probability of each target is required to compute the utilities of the defender and the attacker.

The Bayesian extension to the Stackelberg game allows for multiple types of players, with each associated with its own payoff values [13, 17]. Bayesian games are used to model uncertainty over the payoffs and preferences of the players; indeed more uncertainty can be expressed with increasing number of types. For the security games of interest, there is only one leader type (e.g., only one police force), although there can be multiple follower types (e.g., multiple attacker types trying to infiltrate security). Each follower type is represented using a different payoff matrix, as shown by an example with two attacker types in Table 2.2. The leader does not know the follower's type, but knows the probability distribution over them. The goal is to find the optimal mixed strategy for the leader to commit to, given that the defender could be facing any of the follower types.

2.2.3 Solution Concept: Strong Stackelberg Equilibrium

The solution to a security game is a mixed strategy for the defender that maximizes the expected utility of the defender, given that the attacker learns the mixed strategy of the defender and chooses a best-response for himself. This solution concept is known as a Stackelberg equilibrium [18]. However, the solution concept of choice in all deployed applications is a *strong* form of the Stackelberg equilibrium [19], which assumes that the follower will always break ties in favor of the leader in cases of indifference. This is because a strong Stackelberg equilibrium (SSE) exists in all Stackelberg games, and additionally, the leader can always induce the favorable strong equilibrium by selecting a strategy arbitrarily close to the equilibrium that causes the follower to strictly prefer the desired strategy [20]. Indeed, SSE is the mostly commonly adopted concept in related literature [13, 14, 21].

A SSE for security games is informally defined as follows (the formal definition of SSE is not introduced for brevity, and can instead be found in [16]):

Definition 2.1. A pair of strategies form a *Strong Stackelberg Equilibrium* (SSE) if they satisfy:

1. The defender plays a best-response, that is, the defender cannot get a higher payoff by choosing any other strategy.
2. The attacker play a best-response, that is, given a defender strategy, the attacker cannot get a higher payoff by attacking any other target.
3. The attacker breaks ties in favor of the leader.

2.3 Deployed and Emerging Security Applications

We now talk about five successfully deployed applications that use the concept of strong Stackelberg Equilibrium to suggest security scheduling strategies to the defender in different real-world domains.

2.3.1 ARMOR *for Los Angeles International Airport*

Los Angeles International Airport (LAX) is the largest destination airport in the United States and serves 60–70 million passengers per year. The LAX police use diverse measures to protect the airport, which include vehicular checkpoints, police units patrolling the roads to the terminals, patrolling inside the terminals (with canines), and security screening and bag checks for passengers. The application of game-theoretic approach is focused on two of these measures: (1) placing vehicle checkpoints on inbound roads that service the LAX terminals, including both location and timing, and (2) scheduling patrols for bomb-sniffing canine units at the different LAX terminals. The eight different terminals at LAX have very different characteristics, like physical size, passenger loads, foot traffic or international versus domestic flights. These factors contribute to the differing risk assessments of these eight terminals. Furthermore, the numbers of available vehicle checkpoints and canine units are limited by resource constraints. Thus, it is challenging to optimally allocate these resources to improve their effectiveness while avoiding patterns in the scheduled deployments.

The ARMOR system (Assistant for Randomized Monitoring over Routes) focuses on two of the security measures at LAX (checkpoints and canine patrols) and optimizes security resource allocation using Bayesian Stackelberg games. Take the vehicle checkpoints model as an example. Assume that there are n roads, the police's strategy is placing $m < n$ checkpoints on these roads where m is the maximum number of checkpoints. The adversary may potentially choose to attack through one of these roads. ARMOR models different types of attackers with different payoff functions, representing different capabilities and preferences for the attacker. ARMOR uses DOBSS (Decomposed Optimal Bayesian Stackelberg Solver) [13] to compute the defender's optimal strategy. ARMOR has been successfully deployed since August 2007 at LAX [1, 22].

2.3.2 IRIS *for US Federal Air Marshals Service*

The US Federal Air Marshals Service (FAMS) allocates air marshals to flights originating in and departing from the United States to dissuade potential aggressors and prevent an attack should one occur. Flights are of different importance based on a variety of factors such as the numbers of passengers, the population of source/destination, international flights from different countries, and special events that can change the risks for particular flights at certain times. Security resource allocation in this domain is significantly more challenging than for ARMOR: a limited number of air marshals need to be scheduled to cover thousands of commercial flights each day. Furthermore, these air marshals must be scheduled on tours of flights that obey various constraints (e.g., the time required to board, fly, and disembark). Simply finding schedules for the marshals that meet all of these constraints is a computational challenge. Our task is made more difficult by the need to find a randomized policy that meets these scheduling constraints, while also accounting for the different values of each flight.

Against this background, the IRIS system (Intelligent Randomization In Scheduling) has been developed and has been deployed by FAMS since October 2009 to randomize schedules of air marshals on international flights. In IRIS, the targets are the set of n flights and the attacker could potentially choose to attack one of these flights. The FAMS can assign $m < n$ air marshals that may be assigned to protect these flights.

Since the number of possible schedules exponentially increases with the number of flights and resources, DOBSS is no longer applicable to the FAMS domain. Instead, IRIS uses the much faster ASPEN algorithm [23] to generate the schedule for thousands of commercial flights per day. IRIS also uses an attribute-based preference elicitation system to determine reward values for the Stackelberg game model.

2.3.3 PROTECT *for US Coast Guard*

The US Coast Guard's (USCG) mission includes maritime security of the US coasts, ports, and inland waterways; a security domain that faces increased risks due to threats such as terrorism and drug trafficking. Given a particular port and the variety of critical infrastructure that an attacker may attack within the port, USCG conducts patrols to protect this infrastructure; however, while the attacker has the opportunity to observe patrol patterns, limited security resources imply that USCG patrols

Fig. 2.1 USCG boats patrolling the ports of Boston and NY. (**a**) PROTECT is being used in Boston, (**b**) Extending PROTECT to NY

cannot be at every location 24/7. To assist the USCG in allocating its patrolling resources, the PROTECT (Port Resilience Operational/Tactical Enforcement to Combat Terrorism) model has been designed to enhance maritime security. It has been in use at the port of Boston since April 2011, and now is also in use at the port of New York (Fig. 2.1). Similar to previous applications ARMOR and IRIS, PROTECT uses an attacker–defender Stackelberg game framework, with USCG as the defender against terrorists that conduct surveillance before potentially launching an attack.

The goal of PROTECT is to use game theory to assist the USCG in maximizing its effectiveness in the Ports, Waterways, and Coastal Security (PWCS) Mission. PWCS patrols are focused on protecting critical infrastructure; without the resources to provide 100 percent on scene presence at any, let alone all, of the critical infrastructure, optimization of security resource is critical. Towards that end, unpredictability creates situations of uncertainty for an enemy and can be enough to deem a target less appealing. The PROTECT system, focused on the PWCS patrols, addresses how the USCG should optimally patrol critical infrastructure in a port to maximize protection, knowing that the attacker may conduct surveillance and then launch an attack. While randomizing patrol patterns is key, PROTECT also addresses the fact that the targets are of unequal value, understanding that the attacker will adapt to whatever patrol patterns USCG conducts. The output of PROTECT is a schedule of patrols which includes when the patrols are to begin, what critical infrastructure to visit for each patrol, and what activities to perform at each critical infrastructure.

While PROTECT builds on previous work, it offers some key innovations. First, this system is a departure from the assumption of perfect attacker rationality noted in previous work, relying instead on a quantal response (QR) model [24] of the attacker's behavior. Second, to improve PROTECT's efficiency, a compact representation of the defender's strategy space is used by exploiting equivalence and dominance. Finally, the evaluation of PROTECT for the first time provides real-world data: (a) comparison of human-generated vs PROTECT security schedules, and (b) results from an Adversarial Perspective Team's (human mock attackers) analysis. The PROTECT model is now being extended to the port of New York and it may potentially be extended to other ports in the US.

2.3.4   GUARDS *for US Transportation Security Agency*

The United States Transportation Security Administration (TSA) is tasked with protecting the nation's over 400 airports which services approximately 28,000 commercial flights and up to approximately 87,000 total flights per day. To protect this large transportation network, the TSA employs approximately 48,000 Transportation Security Officers, who are responsible for implementing security activities at each individual airport. While many people are aware of common security activities, such as individual passenger screening, this is just one of many security layers TSA personnel implement to help prevent potential threats [25, 26]. These layers can involve hundreds of heterogeneous security activities executed by limited TSA personnel leading to a complex resource allocation challenge. While activities like passenger screening are performed for every passenger, the TSA cannot possibly run every security activity all the time. Thus, while the resources required for passenger screening are always allocated by the TSA, it must also decide how to appropriately allocate its remaining security officers among the layers of security to protect against a number of potential threats, while facing challenges such as surveillance and an adaptive attacker as mentioned before.

To aid the TSA in scheduling resources to protect airports, a new application called GUARDS (Game-theoretic Unpredictable and Randomly Deployed Security) has been developed. While GUARDS also utilizes Stackelberg games as ARMOR and IRIS, GUARDS faces three key challenges [5]: (1) reasoning about hundreds of heterogeneous security activities; (2) reasoning over diverse potential threats; and (3) developing a system designed for hundreds of end-users. To address those challenges, GUARDS created a new game-theoretic framework that allows for heterogeneous defender activities and compact modeling of a large number of threats and developed an efficient solution technique based on general-purpose Stackelberg game solvers. GUARDS is currently under evaluation and testing for scheduling practices at an undisclosed airport. If successful, the TSA intends to incorporate the system into their unpredictable scheduling practices nationwide.

2.3.5   TRUSTS *for Urban Security in Transit Systems*

In some urban transit systems, including the Los Angeles Metro Rail system, passengers are legally required to purchase tickets before entering but are not physically forced to do so (Fig. 2.2). Instead, security personnel are dynamically deployed throughout the transit system, randomly inspecting passenger tickets.

Fig. 2.2 TRUSTS for transit systems. (**a**) Los Angeles Metro, (**b**) Barrier-free entrance to transit system

This proof-of-payment fare collection method is typically chosen as a more cost-effective alternative to direct fare collection, i.e., when the revenue lost to fare evasion is believed to be less than what it would cost to directly preclude it.

Take the Los Angeles Metro as an example. With approximately 300,000 riders daily, this revenue loss can be significant; the annual cost has been estimated at $5.6 million [27]. The Los Angeles Sheriffs Department (LASD) deploys uniformed patrols on board trains and at stations for fare-checking (and for other purposes such as crime prevention), in order to discourage fare evasion. With limited resources to devote to patrols, it is impossible to cover all locations at all times. The LASD thus requires some mechanism for choosing times and locations for inspections. Any predictable patterns in such a patrol schedule are likely to be observed and exploited by potential fare-evaders. The LASD's current approach relies on humans for scheduling the patrols. However, human schedulers are poor at generating unpredictable schedules; furthermore such scheduling for LASD is a tremendous cognitive burden on the human schedulers who must take into account all of the scheduling complexities (e.g., train timings, switching time between trains, and schedule lengths).

The TRUSTS system (Tactical Randomization for Urban Security in Transit Systems) models the patrolling problem as a leader–follower Stackelberg game [28]. The leader (LASD) pre-commits to a mixed strategy patrol (a probability distribution over all pure strategies), and riders observe this mixed strategy before deciding whether to buy the ticket or not. Both ticket sales and fines issued for fare evasion translate into revenue for the government. Therefore the optimization objective for the leader is to maximize total revenue (total ticket sales plus penalties). Urban transit systems, however, present unique computational challenges since there are exponentially many possible patrol strategies, each subject to both the spatial and temporal constraints of travel within the transit network under consideration. To overcome this challenge, TRUSTS uses a compact representation which captures

Fig. 2.3 The terrorist attacks of 2008 in Mumbai

the spatial as well as temporal structure of the domain. The LASD is currently testing TRUSTS in the LA Metro system by deploying patrols according to the generated schedules and measuring the revenue recovered.

2.3.6 Future Applications

Beyond the deployed and emerging applications above are a number of different application areas. One such area of great importance is securing urban city networks, transportation networks, computer networks and other network centric security domains. For example, after the terrorist attacks in Mumbai of 2008 [29] (refer Fig. 2.3), the Mumbai police have started setting up vehicular checkpoints on roads. We can model the problem faced by the Mumbai police as a security game between the Mumbai police and an attacker. In this urban security game, the pure strategies of the defender correspond to allocations of resources to edges in the network—for example, an allocation of police checkpoints to roads in the city. The pure strategies of the attacker correspond to paths from any *source* node to any *target* node—for example, a path from a landing spot on the coast to the airport.

Another area is protecting forests [30], where we must protect a continuous forest area from extractors by patrols through the forest that seek to deter such extraction activity. With limited resources for performing such patrols, a patrol strategy will seek to distribute the patrols throughout the forest, in space and time, in order to minimize the resulting amount of extraction that occurs or maximize the degree of forest protection. This problem can be formulated as a Stackelberg game and the focus is on computing optimal allocations of patrol density [30].

Another potential application is police patrols for crime suppression which is a data-intensive domain [31]. Thus, it would be promising to use data mining tools on a database of reported crimes and other events to identify the locations to be patrolled, the times at which the game changes, and the types of attackers faced. The idea is to exploit temporal and spatial patterns of crime in the area to be patrolled to determine the priorities on how to use the limited security resources. However, even with all of these applications, we have barely scratched the surface of possibilities in terms of potential applications for multiagent researchers for applying game theory for security.

The Stackelberg game framework can also be applied to adversarial domains that exhibit "contagious" actions for each player. For example, word-of-mouth advertising/viral marketing has been widely studied by marketers trying to understand why one product or video goes "viral" while others go unnoticed [32]. Counter-insurgency is the contest for the support of the local leaders in an armed conflict and can include a variety of operations such as providing security and giving medical supplies. Just as in word-of-mouth advertising and peacekeeping operations, these efforts carry a social effect beyond the action taken that can cause advantageous ripples through the neighboring population. Moreover, multiple intelligent parties attempt to leverage the same social network to spread their message, necessitating an adversary-aware approach to strategy generation. Game-theoretic approaches can be used to generate resource allocations strategies for such large-scale, real world networks. This interaction can be modeled as a graph with one player attempting to spread influence while the other player attempts to stop the probabilistic propagation of that influence by spreading their own influence. This "blocking" problem models situations faced by governments/peacekeepers combatting the spread of terrorist radicalism and armed conflict with daily/weekly/monthy visits with local leaders to provide support and discuss grievances [33].

Game-theoretic methods are also appropriate for modeling resource allocation in cybersecurity [34] such as packet selection and inspection for detecting potential threats in large computer networks [35]. The problem of attacks on computer systems and corporate computer networks gets more pressing each year as the sophistication of the attacks increases together with the cost of their prevention. A number of intrusion detection and monitoring systems are being developed, e.g., deep packet inspection method that periodically selects a subset of packets in a computer network for analysis. However, there is a cost associated with the deep packet inspection, as it leads to significant delays in the throughput of the network. Thus, the monitoring system works under a constraint of limited selection of a fraction of all packets which can be inspected. The attacking/protecting problem can be formulated as a game between two players: the attacker (or the intruder), and the defender (the detection system) [35]. The intruder wants to gain control over (or to disable) a valuable computer in the network by scanning the network, hacking into a more vulnerable system, and/or gaining access to further devices on the computer network. The actions of the attacker can therefore be seen as sending malicious packets from a controlled computer (termed source) to a single or multiple vulnerable computers (termed targets). The objective of the defender is to prevent the intruder from succeeding by selecting the packets for inspection, identifying the attacker, and subsequently thwarting the attack. However, packet inspections cause unwanted latency and hence the defender has to decide where and how to inspect network traffic in order to maximize the probability of a successful malicious packet detection. The computational challenge is efficiently computing the optimal defending strategies for such network scenarios [35].

2.4 Scaling Up To Real-World Problem Sizes

Real world problems, like the FAMS and urban road networks, present billions of pure strategies to both the defender and the attacker. Such large problem instances cannot even be represented in modern computers, let alone solved using previous techniques. We now describe models and algorithms that compute optimal defender strategies for massive real-world security domains. In particular, we describe the following algorithms: (a) ASPEN, an algorithm to compute strong Stackelberg equilibria (SSE) in domains with a *very large* number of pure strategies (up to billions of actions) for the defender [23]; (b) RUGGED, an algorithm to compute the optimal defender strategy in domains with a very large number of pure strategies for both the defender and the attacker [36]; and (c) a hierarchical framework for Bayesian games that is applicable to all Stackelberg solvers, that can be combined with the strategy generation techniques of ASPEN [17]. These algorithms provide scale-ups in real-world domains by efficiently analyzing the strategy space of the players. ASPEN and RUGGED use strategy generation: the algorithms start by considering a minimal set of pure strategies for both the players (defender and attacker). Pure strategies are then generated iteratively, and a strategy is added to the set only if it would help increase the payoff of the corresponding player (a defender's pure strategy is added if it helps increase the defender's payoff). This process is repeated until the optimal solution is obtained. On the other hand, the hierarchical approach pre-processes the Bayesian Stackelberg game and eliminates strategies that can never be the best response of the players. This approach of strategy generation and elimination not only provides the required scale-ups, but also the mathematical guarantees on solution quality. Finally, in this section, we also describe the $d:s$ ratio, an algorithm-independent property of security game instances

---

Algorithm 1: Strategy generation in ASPEN

---

1. Initialize **P**   // initialize, **P**: set of pure strategies of the defender
2. Solve Master Problem   // compute optimal defender mixed strategy, given **P**
3. Calculate cost coefficients from solution of master
4. Update objective of slave problem with coefficients
         // compute whether a pure strategy will increase defender's payoff
5. Solve Slave Problem   // generate the pure strategy that is likely to increase
the defender's payoff the most
if Optimal solution obtained **then**
   6. Return (**x**, **P**)
else
   7. Extract new pure strategy and add to **P**
   8. Repeat from Step 2

---

that influences the hardness of computation for the given problem instance and is robust across domains, domain representations, algorithms and underlying solvers.

2.4.1 Scaling Up with Defender Pure Strategies

In this section, we describe how ASPEN generates pure strategies for the defender in domains where the number of pure strategies of the defender can be prohibitively large. As an example, let us consider the problem faced by the Federal Air Marshals Service (FAMS). There are currently tens of thousands of commercial flights flying each day, and public estimates state that there are thousands of air marshals that are scheduled daily by the FAMS [37]. Air marshals must be scheduled on tours of flights that obey logistical constraints (e.g., the time required to board, fly, and disembark). An example of a valid schedule is an air marshal assigned to a round trip tour from Los Angeles to New York and back.

ASPEN [23] casts this problem as a security game, where the attacker can choose any of the flights to attack, and each air marshal can cover one schedule. Each schedule here is a feasible set of targets that can be covered together; for the FAMS, each schedule would represent a flight tour which satisfies all the logistical constraints that an air marshal could fly. A *joint schedule* then would assign every air marshal to a flight tour, and there could be exponentially many joint schedules in the domain. A pure strategy for the defender in this security game is a joint schedule. As mentioned previously, ASPEN employs strategy generation since all the defender pure strategies (or joint schedules) cannot be enumerated for such a massive problem. ASPEN decomposes the problem into a *master* problem and a *slave* problem, which are then solved iteratively, as described in Algorithm 1. Given a limited number of pure strategies, the master solves for the defender and the attacker optimization constraints, while the slave is used to generate a new pure strategy for the defender in every iteration.

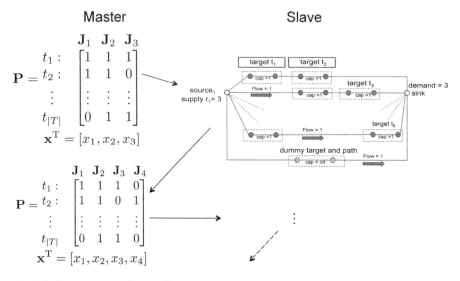

Fig. 2.4 Strategy generation employed in ASPEN: The schedules for a defender are generated iteratively. The *slave* problem is a novel minimum-cost integer flow formulation that computes the new pure strategy to be added to **P**; J_4 is computed and added in this example

The iteratively process of Algorithm 1 is graphically depicted in Fig. 2.4. The master operates on the pure strategies (joint schedules) generated thus far (Step 2), which are represented using the matrix **P**. Each column of **P**, J_j, is one pure strategy (or joint schedule). An entry P_{ij} in the matrix **P** is 1 if a target t_i is covered by joint-schedule J_j, and 0 otherwise. The objective of the master problem is to compute **x**, the optimal mixed strategy of the defender over the pure strategies in **P**. The objective of the slave problem is to generate the best joint schedule to add to **P** (Step 5). The best joint schedule is identified using the concept of *reduced costs* [38] (Step 3–4), which measures if a pure strategy can potentially increase the defender's expected utility (the details of the approach are provided in [23]). While a naïve approach would be to iterate over all possible pure strategies to identify the pure strategy with the maximum potential, ASPEN uses a novel minimum-cost integer flow problem to efficiently identify the best pure strategy to add. ASPEN always converges on the optimal mixed strategy for the defender; the proof can be found in [23].

Employing strategy generation for large optimization problems is not an "out-of-the-box" approach, the problem has to be formulated in a way that allows for domain properties to be exploited. The novel contribution of ASPEN is to provide a linear formulation for the master and a minimum-cost integer flow formulation for the slave, which enable the application of strategy generation techniques. Additionally, ASPEN also provides a branch-and-bound heuristic to reason over attacker actions. This branch-and-bound heuristic provides a further order of magnitude speed-up, allowing ASPEN to handle the massive sizes of real-world problems.

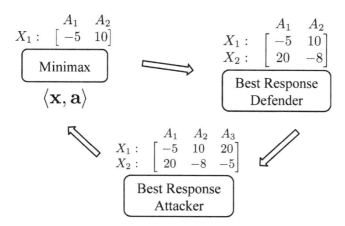

Fig. 2.5 Strategy Generation employed in RUGGED: The pure strategies for both the defender and the attacker are generated iteratively

2.4.2 Scaling Up with Defender and Attacker Pure Strategies

In this section, we describe how RUGGED generates pure strategies for both the defender and the attacker in domains where the number of pure strategies of the players are exponentially large. Let us consider as an example the urban network security game.

Here, strategy generation is required for both the defender and the attacker since the number of pure strategies of both the players are prohibitively large. Figure 2.5 shows the working of RUGGED: here, the minimax module generates the optimal mixed strategies $\langle \mathbf{x}, \mathbf{a} \rangle$ for the two players, whereas the two best response modules generate new strategies for the two players respectively. The rows X_i in the figure are the pure strategies for the defender, they would correspond to an allocation of checkpoints in the urban road network domain. Similarly, the columns A_j are the pure strategies for the attacker, they represent the attack paths in the urban road network domain. The values in the matrix represent the payoffs to the defender. RUGGED models the domain as a zero-sum game, and computes the minimax equilibrium, since the payoff of the minimax strategy is equivalent to the SSE payoff in zero-sum games [39].

The contribution of RUGGED is to provide the mixed integer formulations for the best response modules which enable the application of such a strategy generation approach. RUGGED can compute the optimal solution for deploying up to 4 resources in real-city network with as many as 250 nodes within a reasonable time frame of 10 h (the complexity of this problem can be estimated by observing that both the best response problems are NP-hard themselves [36]). While enhancements to RUGGED are required for deployment in larger real-world domains, RUGGED has opened new possibilities for scaling up to larger games.

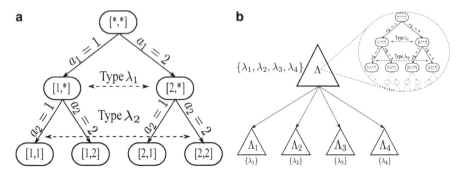

Fig. 2.6 Hierarchical approach for solving Bayesian Stackelberg games. (**a**) Attacker action tree, for a Bayesian game with attackers of two types, with two attacker actions each, (**b**) Hierarchical game tree, decomposing a Bayesian Stackelberg game with 4 types into 4 *restricted* games with one type each

2.4.3 Scaling Up with Attacker Types

The different preferences of different attacker types are modeled through Bayesian Stackelberg games. Computing the optimal leader strategy in Bayesian Stackelberg game is NP-hard [14], and polynomial time algorithms cannot achieve approximation ratios better than $O(types)$ [40]. We now describe HBGS, a hierarchical technique for solving large Bayesian Stackelberg games that decomposes the entire game into many hierarchically-organized *restricted* Bayesian Stackelberg games; it then utilizes the solutions of these restricted games to more efficiently solve the larger Bayesian Stackelberg game [17].

Figure 2.6a shows the *attacker action tree* (a tree depicting all possible pure strategies of an attacker in a Bayesian game, arranged as a tree) of an example Bayesian Stackelberg game with 2 types and 2 actions per attacker type. The leaf nodes of this tree are all the possible action combinations for the attacker in this game, for example, leaf $[1, 1]$ implies that the attackers of both types λ_1 and λ_2 chose action a_1. All the leaves of this tree (i.e. all combinations of attacker strategies for all attacker types) need to be evaluated before the optimal strategy for the defender can be computed. This tree is typically evaluated using branch-and-bound. This requirement to evaluate the exponential number of leaves (or pure strategy combinations) is the cause of NP-hardness of Bayesian Stackelberg games; indeed the performance of algorithms can be improved if leaves can be pruned by pre-processing.

The overarching idea of hierarchical structure is to improve the performance of branch-and-bound on the attacker action tree (Fig. 2.6a) by pruning leaves of this tree. It decomposes the Bayesian Stackelberg game into many hierarchically-organized smaller games, as shown by an example in Fig. 2.6b. Each of the restricted games ("child" nodes in Fig. 2.6b) consider only a few attacker types, and are thus exponentially smaller that the Bayesian Stackelberg game at the "parent".

For instance, in this example, the Bayesian Stackelberg game has 4 types ($\langle \lambda_1, \lambda_2, \lambda_3, \lambda_4 \rangle$) and it is decomposed into 4 *restricted* games (leaf nodes) with each restricted game having exactly 1 attacker type. The solutions obtained for the restricted games at the child nodes of the hierarchical game tree are used to provide: (a) pruning rules, (b) tighter bounds, and (c) efficient branching heuristics to solve the bigger game at the parent node faster.

Such hierarchical techniques have seen little application towards obtaining optimal solutions in Bayesian games, while Stackelberg settings have not seen any application of such hierarchical decomposition. This hierarchical idea of HBGS has also been combined with the strategy generation of ASPEN by the HBSA algorithm [17], which is now opening new possibilities to handle uncertainty in large games.

2.4.4 Characterizing Hard Security Game Problem Instances

In this section, we will focus on the runtime required by the different algorithms that compute solutions for instances of security games for three of the domains described above: the LAX or ARMOR domain, the FAMS or IRIS domain and the Coast Guard or PROTECT domain. We will describe the *d:s* ratio, a domain-spanning measure of the density of defender coverage in any security problem. We will then present results that show that the computationally hardest problem instances of security games occur when this ratio is 0.5.

However, first lets look at the runtime required by the different algorithms that compute solutions for different security domains. Figure 2.7 shows the runtime for computing solutions using DOBSS, ASPEN and a multiple-attacker branch-and-price algorithm for the ARMOR, IRIS and PROTECT[1] domains respectively. The *x*-axis in each figure shows the number of available resources to the defender, and the *y*-axis shows the runtime in seconds. In the ARMOR and the IRIS domains, we define the number of resources as the number of officers available to the defender; in the PROTECT domain, we define it as the maximum feasible tour length. These graphs show that there is no unified value of the number of defender resources which makes security game instances hard to solve. Even normalizing the number of resources by the number of targets shows similar inconsistencies across different domains.

We now describe the concept of the *deployment-to-saturation* (*d:s*) ratio [41], a concept that unifies the domain independent properties of problem instances across different security domains. Specifically, this concept of *d:s* ratio has been applied to the three security domains described above, eight different MIP algorithms

---

[1]The deployed application of PROTECT uses a quantal response model whereas the multiple attacker branch-and-price algorithm computes SSE as originally defined [41]. Here, by the PROTECT domain, we refer to the security domain where defender has to execute patrols that are bounded by their tour length.

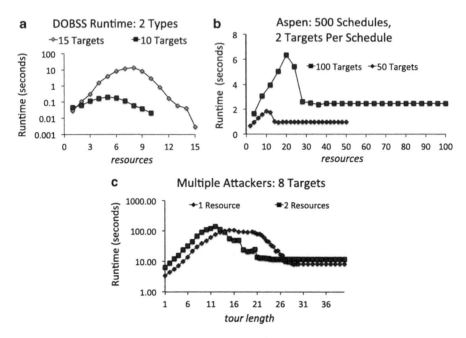

Fig. 2.7 Average running time of DOBSS, ASPEN and multiple-attacker branch-and-price algorithm for SPNSC, SPARS and SPPC domains respectively

(introduced in literature for these three domains [41]), five different underlying MIP solvers, two different equilibrium concepts in Stackelberg security games (ε-SSE [42] being the second equilibrium concept), and a variety of domain sizes and conditions.

More specifically, the deployment-to-saturation (d:s) ratio is defined in terms of *defender resources*, a concept whose precise definition differs from one domain to another. Given this definition, *deployment* denotes the number of defender resources available to be allocated, and *saturation* denotes the minimum number of defender resources such that the addition of further resources beyond this point yields no increase in the defender's expected utility. As mentioned earlier, for the ARMOR and the IRIS domains, deployment denotes the number of available security personnel, whereas saturation refers to the minimum number of officers required to cover all the targets with a probability of 1. For the PROTECT domain, deployment denotes the maximum feasible tour length, while saturation denotes the minimum tour length required by the team of defender resources such that the team can tour all the targets with a probability of 1. Sample results for these three domains are shown in Fig. 2.8, where the x-axis is the d:s ratio and the primary y-axis is runtime in seconds. The secondary y-axis and the dashed lines show a *phase transition* that we describe later. The important deduction to draw from these graphs is that the hardest computation is required when the d:s ratio is close to 0.5.

There are two important implications of this finding. First, new algorithms should be compared on the hardest problem instances; indeed, most previous

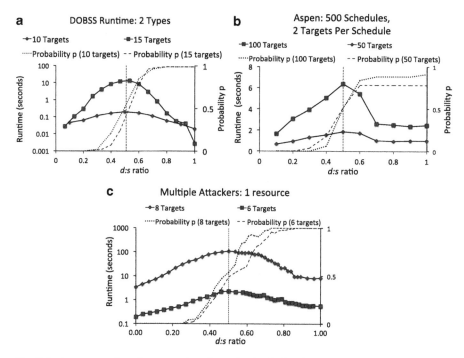

Fig. 2.8 Average runtime for computing the optimal solution for an example setting for all the three security domains, along with the probability p plotted on the secondary y-axis. The vertical dotted line shows $d:s = 0.5$

research has compared the runtime performance of algorithms only at low $d:s$ ratios, where problems are comparatively easy [41]. Second, this computationally hard region is the point where optimization offers the greatest benefit to security agencies, implying that problems in this region deserve increased attention from researchers [41]. Furthermore, we present an analysis of the runtime results using the concept of a *phase transition* in the decision version of SSE optimization problem.

All the runtime results show an easy-hard-easy computational pattern as the $d:s$ ratio increases from 0 to 1, with the hardest problems at $d:s = 0.5$. Such easy–hard–easy patterns have also been observed in other NP-complete problems, most notably 3-SAT [43, 44]. In 3-SAT, the hardness of the problems varies with the clause-to-variable (c/v) ratio, with the hardest instances occurring at about $c/v = 4.26$. The SAT community has used the concept of phase transitions to better understand this hardness peak. Phase transitions are defined for decision problems; and the decision version $SSE(D)$ of the SSE optimization problem asks whether there exists a defender strategy that guarantees expected utility of at least the given value D.

The results plotting the phase transition are also shown in Fig. 2.8. The x-axis shows the $d:s$ ratio, the primary y-axis shows the runtime in seconds, and the secondary y-axis shows the probability p of finding a solution to $SSE(D^*)$. D^* was chosen by computing the median defender utility for 100 random problem instances for the domain [41]. The runtimes are plotted using solid lines, and p is plotted using a dashed line. Figure 2.8a presents results for the DOBSS algorithm [13] for the ARMOR domain for 10 targets and 2 and 3 attacker types. As expected, the d:s ratio of 0.5 corresponds with $p = 0.51$ as well as the computationally hardest instances; more interestingly, p undergoes a phase transition as the d:s grows. Similarly, Fig. 2.8b shows results for the ASPEN algorithm [23] for the IRIS domain with 500 schedules, 2 targets per schedule and 50 and 100 schedules, and Fig. 2.8c shows results for the multiple attacker PROTECT domain [41] for 8 targets. In both cases, we again observe a phase transition in p.

2.5 Open Research Issues

While the deployed applications have advanced the state of the art, significant future research remains to be done. In the following, we highlight some key research challenges, including scalability, robustness, human adversary modeling, and mixed-initiative optimization. The main point we want to make is that this research does not require access to classified information of any kind. Problems, solution approaches and datasets are well specified in the papers discussed below,

Scalability: The first research challenge is improving the scalability of our algorithms for solving Stackelberg (security) games. The strategy space of both the defender and the attacker in these games may exponentially increase with the number of security activities, attacks, and resources. As we scale up to larger domains, it is critical to develop newer algorithms that scale up significantly beyond the limits of the current state of the art of Bayesian Stackelberg solvers. Driven by the growing complexity of applications, a sequence of algorithms for solving security games have been developed including DOBSS [13], ERASER [16], ASPEN [23], HBGS [17] and RUGGED [36]. However, existing algorithms still cannot scale up to very large scale domains such as scheduling randomized checkpoints in cities (while RUGGED computes optimal solutions much faster than any of the previous approaches, much work remains to be done for it to be applicable on a large urban road network).

Robustness: The second challenge is improving solutions' robustness. Classical game theory solution concepts often make assumptions on the knowledge, rationality, and capability (e.g., perfect recall) of players. Unfortunately, these assumptions could be wrong in real-world scenarios. Therefore, while computing the defender's optimal strategy, algorithms should take into account various uncertainties faced in the domain, including payoff noise [45], execution/observation error [46], uncertain capability [47]. While there are algorithms for dealing with different types of uncertainties, there is no general algorithm/framework that can deal with different types of uncertainty simultaneously. Furthermore, existing work assumes that the attacker knows (or with a small noise) the defender's strategy and there is no formal framework to model the attacker's belief update process and how it makes tradeoffs in consideration of surveillance cost, which remains an open issue for in future research.

One required research direction with respect to robustness is addressing bounded rationality of human adversaries, which is a fundamental problem that can affect the performance of our game theoretic solutions. Recently, there has been some research on applying ideas (e.g., prospect theory [48], and quantal response [24]) from social science or behavioral game theory within security game algorithms [42, 49]. Previous work usually applies existing frameworks and sets the parameters of these frameworks by experimental tuning or learning. However, in real-world security domains, we may have very limited data, or may only have some limited information on the biases displayed by adversaries. It is thus still a challenging problem to build high fidelity human attacker models that can address human bounded rationality. Furthermore, since real-world human attackers are sometimes distributed coalitions of socially, culturally and cognitively-biased agents, acting behind a veil of uncertainty, we may need significant interdisciplinary research to build in social, cultural and coalitional biases into our adversary models.

Mixed-Initiative Optimization: Another challenging research problem in security games is mixed-initiative optimization in which human users and software assistants collaborate to make security decisions [50]. There often exist different types of constraints in security applications. For instance, the defender always has resource constraints, e.g., the numbers of available vehicle checkpoints, canine units, or air marshals. In addition, human users may place constraints on the defender's actions to affect the output of the game when they are faced with exceptional circumstances and extra knowledge. For instance, in the ARMOR system there could be forced checkpoints (e.g., when the Governor is flying) and forbidden checkpoints. Existing applications simply compute the optimal solution to meet all the constraints (if possible). Unfortunately, these user defined constraints may lead to poor (or infeasible) solutions due to the users' bounded rationality and insufficient information about how constraints affect the solution quality. Significantly better solution quality can be obtained if some of these constraints can be relaxed. However, there may be infinitely many ways of relaxing constraints and the software assistant may not know which constraints can be relaxed and by how much, as well as the real-world consequences of relaxing some constraints.

Thus, it is promising to adopt a mixed-initiative approach in which human users and software assistants collaborate to make security decisions. However, designing an efficient mixed-initiative optimization approach is not trivial and there are five major challenges. First, the scale of security games and constraints prevent us from using an exhaustive search algorithm to explore all constraint sets. Second, the user's incomplete information regarding the consequences of relaxing constraints requires preference elicitation support. Third, the decision making of shifting control between the user and the software assistant is challenging. Fourth, it is difficult to evaluate the performance of a mixed-initiative approach. Finally, it is a challenging problem to design good user interfaces for the software assistant to explain how constraints affect the solution quality. What remains to be done for the mixed-initiative approach includes sensitivity analysis for understanding how different constraints affect the solution quality, inference/learning for discovering directions of relaxing constraints, search for finding constraint sets to explore, preference elicitation for finding the human user's preference of different constraint sets, and interface design for explaining the game theoretic solver's performance.

Multi-objective Optimization: In existing applications such as ARMOR, IRIS and PROTECT, the defender is trying to maximize a single objective. However, there are domains where the defender has to consider multiple objectives simultaneously. For example, the Los Angeles Sheriff's Department (LASD) needs to protect the city's metro system from ticketless travelers, common criminals, and terrorists. From the perspective of LASD, each one of these attacker types provides a unique threat (lost revenue, property theft, and loss of life). Given this diverse set of threats, selecting a security strategy is a significant challenge as no single strategy can minimize the threat for all attacker types. Thus, tradeoffs must be made and protecting more against one threat may increase the vulnerability to another threat. However, it is not clear how LASD should weigh these threats when determining the security strategy to use. One could attempt to establish methods for converting the different threats into a single metric. However, this process can become convoluted when attempting to compare abstract notions such as safety and security with concrete concepts such as ticket revenue.

Multi-objective security games (MOSG) have been proposed to address the challenges of domains with multiple incomparable objectives [51]. In an MOSG, the threats posed by the attacker types are treated as different objective functions which are not aggregated, thus eliminating the need for a probability distribution over attacker types. Unlike Bayesian security games which have a single optimal solution, MOSGs have a set of pareto-optimal (non-dominated) solutions which is referred to as the Pareto frontier. By presenting the pareto frontier to the end user, they are able to better understand the structure of their problem as well as the trade-offs between different security strategies. As a result, end users are able to make a more informed decision on which strategy to enact. Existing approaches so far assume that each attacker type has a single objective and there is no uncertainty regarding each attacker type's payoffs. It is challenging to develop algorithms for solving multi-objective security games with multiple attacker objectives and uncertain attacker payoffs.

In addition to the above research challenges, there are other on-going challenges such as preference elicitation for acquiring necessary domain knowledge in order to build game models and evaluation of the game theoretic applications [52].

Acknowledgements This research is supported by MURI grant W911NF-11-1-0332, ONR grant N00014-08-1-0733 and by the United States Department of Homeland Security through the Center for Risk and Economic Analysis of Terrorism Events (CREATE) under grant number 2010-ST-061-RE0001. All opinions, findings, conclusions and recommendations in this document are those of the authors and do not necessarily reflect views of the United States Department of Homeland Security.

References

1. Pita, J., Jain, M., Western, C., Portway, C., Tambe, M., Ordonez, F., Kraus, S., Parachuri, P.: Deployed ARMOR protection: The Application of a Game-Theoretic Model for Security at the Los Angeles International Airport. In: Proc. of The 7th International Conference on Autonomous Agents and Multiagent Systems (AAMAS). (2008) 125–132
2. Tsai, J., Rathi, S., Kiekintveld, C., Ordonez, F., Tambe, M.: IRIS: a Tool for Strategic Security Allocation in Transportation Networks. In: Proc. of The 8th International Conference on Autonomous Agents and Multiagent Systems (AAMAS). (2009) 37–44
3. An, B., Pita, J., Shieh, E., Tambe, M., Kiekintveld, C., Marecki, J.: GUARDS and PROTECT: Next Generation Applications of Security Games. SIGECOM **10** (March 2011) 31–34
4. Shieh, E., An, B., Yang, R., Tambe, M., Baldwin, C., DiRenzo, J., Maule, B., Meyer, G.: PROTECT: A Deployed Game Theoretic System to Protect the Ports of the United States. In: Proc. of The 11th International Conference on Autonomous Agents and Multiagent Systems (AAMAS). (2012)
5. Pita, J., Tambe, M., Kiekintveld, C., Cullen, S., Steigerwald, E.: GUARDS - Game Theoretic Security Allocation on a National Scale. In: Proc. of The 10th International Conference on Autonomous Agents and Multiagent Systems (AAMAS). (2011)
6. Yin, Z., Jiang, A., Johnson, M., Tambe, M., Kiekintveld, C., Leyton-Brown, K., Sandholm, T., Sullivan, J.: TRUSTS: Scheduling Randomized Patrols for Fare Inspection in Transit Systems. In: Proc. of The 24th Conference on Innovative Applications of Artificial Intelligence (IAAI). (2012)
7. von Stackelberg, H.: Marktform und Gleichgewicht. Springer, Vienna (1934)
8. Gatti, N.: Game Theoretical Insights in Strategic Patrolling: Model and Algorithm in Normal-Form. In: ECAI-08. (2008) 403–407
9. Lye, K., Wing, J.M.: Game Strategies in Network Security. International Journal of Information Security **4**(1–2) (2005) 71–86
10. Brown, G., Carlyle, M., Kline, J., Wood, K.: A Two-Sided Optimization for Theater Ballistic Missile Defense. In: Operations Research. Volume 53. (2005) 263–275
11. Sandler, T., M., D.G.A.: Terrorism and Game Theory. Simulation and Gaming **34**(3) (2003) 319–337
12. Avenhaus, R., von Stengel, B., Zamir, S.: Inspection Games. In Aumann, R.J., Hart, S., eds.: Handbook of Game Theory. Volume 3. North-Holland, Amsterdam (2002) 1947–1987
13. Paruchuri, P., Pearce, J.P., Marecki, J., Tambe, M., Ordonez, F., Kraus, S.: Playing Games with Security: An Efficient Exact Algorithm for Bayesian Stackelberg Games. In: Proc. of The 7th International Conference on Autonomous Agents and Multiagent Systems (AAMAS). (2008) 895–902
14. Conitzer, V., Sandholm, T.: Computing the Optimal Strategy to Commit to. In: Proc. of the ACM Conference on Electronic Commerce (ACM-EC). (2006) 82–90
15. Brown, G., Carlyle, M., Salmeron, J., Wood, K.: Defending Critical Infrastructure. In: Interfaces. Volume 36. (2006) 530 – 544
16. Kiekintveld, C., Jain, M., Tsai, J., Pita, J., Tambe, M., Ordonez, F.: Computing Optimal Randomized Resource Allocations for Massive Security Games. In: Proc. of The 8th International Conference on Autonomous Agents and Multiagent Systems (AAMAS). (2009) 689–696
17. Jain, M., Kiekintveld, C., Tambe, M.: Quality-Bounded Solutions for Finite Bayesian Stackelberg Games: Scaling Up. In: Proc. of The 10th International Conference on Autonomous Agents and Multiagent Systems (AAMAS). (2011)
18. Leitmann, G.: On Generalized Stackelberg Strategies. Optimization Theory and Applications **26**(4) (1978) 637–643
19. Breton, M., Alg, A., Haurie, A.: Sequential stackelberg equilibria in two-person games. Optimization Theory and Applications **59**(1) (1988) 71–97
20. von Stengel, B., Zamir, S.: Leadership with Commitment to Mixed Strategies. Technical Report LSE-CDAM-2004-01, CDAM Research Report (2004)

21. Osbourne, M.J., Rubinstein, A.: A Course in Game Theory. MIT Press (1994)
22. Jain, M., Tsai, J., Pita, J., Kiekintveld, C., Rathi, S., Tambe, M., Ordonez, F.: Software Assistants for Randomized Patrol Planning for the LAX Airport Police and the Federal Air Marshal Service. Interfaces **40** (2010) 267–290
23. Jain, M., Kardes, E., Kiekintveld, C., Ordonez, F., Tambe, M.: Security Games with Arbitrary Schedules: A Branch and Price Approach. In: Proc. of The 24th AAAI Conference on Artificial Intelligence. (2010) 792–797
24. McKelvey, R.D., Palfrey, T.R.: Quantal Response Equilibria for Normal Form Games. Games and Economic Behavior **10**(1) (1995) 6–38
25. TSA: Layers of Security: What We Do. http://www.tsa.gov/what_we_do/layers/index.shtm (2011)
26. TSA: Transportation Security Administration — U.S. Department of Homeland Security. http://www.tsa.gov/ (2011)
27. Hamilton, B.A.: Faregating Analysis. Report Commissioned by the LA Metro. http://boardarchives.metro.net/Items/2007/11_November/20071115EMACItem27.pdf (2007)
28. Jiang, A.X., Yin, Z., Kietkintveld, C., Leyton-Brown, K., Sandholm, T., Tambe, M.: Towards Optimal Patrol Strategies for Urban Security in Transit Systems. In: Proc. of the AAAI Spring Symposium on Game Theory for Security, Sustainability and Health. (2012)
29. Chandran, R., Beitchman, G.: Battle for Mumbai Ends, Death Toll Rises to 195. Times of India (29 November 2008) http://articles.timesofindia.indiatimes.com/2008-11-29/india/27930171_1_taj-hotel-three-terrorists-nariman-house.
30. Johnson, M., Fang, F., Yang, R., Tambe, M., Albers, H.: Patrolling to Maximize Pristine Forest Area. In: Proc. of the AAAI Spring Symposium on Game Theory for Security, Sustainability and Health. (2012)
31. Ordonez, F., Tambe, M., Jara, J.F., Jain, M., Kiekintveld, C., Tsai, J.: Deployed Security Games for Patrol Planning. In: Handbook on Operations Research for Homeland Security. (2008)
32. Trusov, M., Bucklin, R.E., Pauwels, K.: Effects of Word-of-Mouth versus Traditional Marketing: Findings from an Internet Social Networking Site. Journal of Marketing **73** (2009)
33. Howard, N.J.: Finding Optimal Strategies for Influencing Social Networks in Two Player Games. Master's thesis, MIT, Sloan School of Management (2011)
34. Alpcan, T.: Network Security: A Decision and Game-Theoretic Approach. Cambridge University Press (2010)
35. Vanek, O., Yin, Z., Jain, M., Bosansky, B., Tambe, M., Pechoucek, M.: Game-Theoretic Resource Allocation for Malicious Packet Detection in Computer Networks. In: Proc. of The 11th International Conference on Autonomous Agents and Multiagent Systems (AAMAS). (2012)
36. Jain, M., Korzhyk, D., Vanek, O., Pechoucek, M., Conitzer, V., Tambe, M.: A Double Oracle Algorithm for Zero-Sum SEcurity games on Graphs. In: Proc. of The 10th International Conference on Autonomous Agents and Multiagent Systems (AAMAS). (2011)
37. Keteyian, A.: TSA: Federal Air Marshals. (2010) http://www.cbsnews.com/stories/2010/02/01/earlyshow/main6162291.shtml, *retrieved* Feb 1, 2011.
38. Bertsimas, D., Tsitsiklis, J.N.: Introduction to Linear Optimization. Athena Scientific (1994)
39. Yin, Z., Korzhyk, D., Kiekintveld, C., Conitzer, V., Tambe, M.: Stackelberg vs. Nash in security games: interchangeability, equivalence, and uniqueness. In: AAMAS. (2010) 1139–1146
40. Letchford, J., Conitzer, V., Munagala, K.: Learning and Approximating the Optimal Strategy to Commit To. In: Second International Symposium on Algorithmic Game Theory (SAGT). (2009) 250–262
41. Jain, M., Leyton-Brown, K., Tambe, M.: The Deployment-to-Saturation Ratio in Security Games. In: Proc. of The 26th AAAI Conference on Artificial Intelligence (AAAI). (2012)
42. Pita, J., Jain, M., Tambe, M., Ordóñez, F., Kraus, S.: Robust Solutions to Stackelberg Games: Addressing Bounded Rationality and Limited Observations in Human Cognition. Artificial Intelligence **174**(15) (2010) 1142–1171
43. Cheeseman, P., Kanefsky, B., Taylor, W.M.: Where the Really Hard Problems are. In: Proceedings of the International Joint Conference on Artificial Intelligence. (1991) 331–337

44. Mitchell, D., Selman, B., Levesque, H.: Hard and Easy Distributions of SAT Problems. In: Proceedings of the American Association for Artificial Intelligence. (1992) 459–465
45. Kiekintveld, C., Marecki, J., Tambe, M.: Approximation Methods for Infinite Bayesian Stackelberg Games: Modeling Distributional Uncertainty. In: Proc. of The 10th International Conference on Autonomous Agents and Multiagent Systems (AAMAS). (2011)
46. Yin, Z., Jain, M., Tambe, M., Ordonez, F.: Risk-Averse Strategies for Security Games with Execution and Observational Uncertainty. In: Proc. of The 25th AAAI Conference on Artificial Intelligence (AAAI). (2011) 758–763
47. An, B., Tambe, M., Ordonez, F., Shieh, E., Kiekintveld, C.: Refinement of Strong Stackelberg Equilibria in Security Games. In: Proc. of the 25th Conference on Artificial Intelligence. (2011) 587–593
48. Kahneman, D., Tvesky, A.: Prospect Theory: An Analysis of Decision Under Risk. Econometrica **47**(2) (1979) 263–291
49. Yang, R., Kiekintveld, C., Ordonez, F., Tambe, M., John, R.: Improving Resource Allocation Strategy Against Human Adversaries in Security Games. In: IJCAI. (2011)
50. An, B., Jain, M., Tambe, M., Kiekintveld, C.: Mixed-Initiative Optimization in Security Games: A Preliminary Report. In: Proc. of the AAAI Spring Symposium on Help Me Help You: Bridging the Gaps in Human-Agent Collaboration. (2011) 8–11
51. Brown, M., An, B., Kiekintveld, C., Ordonez, F., Tambe, M.: Multi-objective optimization for security games. In: Proc. of The 11th International Conference on Autonomous Agents and Multiagent Systems (AAMAS). (2012)
52. Taylor, M.E., Kiekintveld, C., Western, C., Tambe, M.: A Framework For Evaluating Deployed Security Systems: Is There a Chink in your ARMOR? Informatica **34** (2010) 129–139

Chapter 3
Adversarial Dynamics: The Conficker Case Study

Daniel Bilar, George Cybenko, and John Murphy

Abstract It is well known that computer and network security is an adversarial challenge. Attackers develop exploits and defenders respond to them through updates, service packs or other defensive measures. In non-adversarial situations, such as automobile safety, advances on one side are not countered by the other side and so progress can be demonstrated over time. In adversarial situations, advances by one side are countered by the other and so oscillatory performance typically emerges. This paper contains a detailed study of the coevolution of the Conficker Worm and associated defenses against it. It demonstrates, in concrete terms, that attackers and defenders each present *moving targets* to the other. After detailing specific adaptations of attackers and defenders in the context of Conficker and its variants, we briefly develop a quantitative model for explaining the coevolution based on what we call *Quantitative Attack Graphs* (QAG) which involve attackers selecting shortest paths through an attack graph with defenders investing in hardening the shortest path edges appropriately.

3.1 Introduction

Progress in operational cyber security has been difficult to demonstrate. In spite of the research and development investments made over more than 30 years, many government, commercial and consumer information systems continue to be

D. Bilar (✉) • J. Murphy
Process Query Systems LLC, 16 Cavendish Ct., Lebanon NH 03766 Siege Technologies,
33 S Commercial St., Manchester NH 03101, USA
e-mail: dbilar@acm.org; jmurphy@flowtraq.com

G. Cybenko
Process Query Systems LLC, 16 Cavendish Ct., Lebanon NH 03766 Dartmouth College,
Hanover NH 03750, USA
e-mail: gvc@flowtraq.com; gvc@dartmouth.edu

S. Jajodia et al. (eds.), *Moving Target Defense II: Application of Game Theory and Adversarial Modeling*, Advances in Information Security 100, DOI 10.1007/978-1-4614-5416-8_3, © Springer Science+Business Media New York 2013

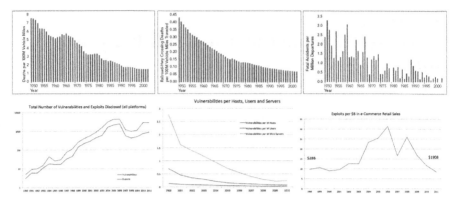

Fig. 3.1 *Top Left*—TRAFFIC FATALITY RATES: U.S. Motor Vehicle Fatalities per 100 Million Vehicle Miles, 1950–2003. (Source: National Highway Traffic Safety Administration). *Top Middle*—RAILROAD-HIGHWAY CROSSING FATALITY RATES: U.S. Railroad–Highway Crossing Fatalities per 100 Million Vehicle Miles, 1950–2003. (Source: Federal Railroad Administration, Bureau of Transportation Statistics). *Top Right*—AVIATION FATALITY RATES: Fatal accidents per million departures for U.S. scheduled service airlines, 1950–2003. Accidents due to sabotage or terrorism are not included. (Source: Air Transport Association). *Bottom Left*—VULNERABILITY AND EXPLOIT COUNTS: Total Number of Vulnerabilities and Exploits (Graph produced by P. Sweeney from data in OSVDB). *Bottom Middle*—VULNERABILITIES PER HOSTS/USERS/SERVERS: Total Number of Vulnerabilities normalized by internet hosts, users and servers (Graph produced by P. Sweeney from data in OSVDB, Internet Systems Consortium (number of hosts on the Internet), Netcraft webserver survey (number of webservers on the Internet), and Internet World Stats (number of Internet users).). *Bottom Right*—EXPLOITS NORMALIZED BY ECOMMERCE: Total number of exploits per billion dollars of e-commerce (Graph produced by P. Sweeney from data in OSVDB and www.census.gov/estats)

successfully attacked and exploited on a routine basis. By contrast, research and development investments in automobile, rail and aviation safety over the same time periods have led to significant, demonstrable improvements in the corresponding domains.

Advances in standard performance measures for automobile, train and airline transportation (namely fatalities per unit of travel) are depicted at the top of Fig. 3.1, while corresponding measures for cyber security are depicted at the bottom.

A major difference between automobile safety and information security is that in the former the adversaries are natural laws that don't change, while in the latter the adversaries are rational humans who adapt quickly and creatively. Consequently, we argue that we cannot understand or model the cyber security landscape in terms of steadily making progress towards an asymptotic "solution" as the transportation statistics suggest is happening in that domain.

In Fig. 3.1, the bottom row of plots show, from left to right, vulnerabilities and/or exploits in absolute numbers (bottom left), normalized by the estimated number of hosts, users and servers (bottom middle) and normalized by the estimated amount of e-commerce (bottom right). The total number of reported vulnerabilities and exploits is growing at first and leveling off (note that this is a logarithmic plot), with some noticeable oscillations especially in recent years; however, the trend is most meaningful if normalized by some measure of corresponding "activity." For example, traffic fatality statistics are routinely normalized by vehicle miles travelled or aircraft takeoffs.

We have not settled the matter about what the correct or analogous normalization for cyber vulnerabilities and exploits would be. If we were to normalize by the number of different operating system platforms for example, the plot would basically resemble the bottom left because there simply are not that many different platforms available in the market. If we normalize by the total number of users, hosts or servers, there is a precipitous drop in vulnerabilities as the bottom middle plot shows. However, if we normalize by e-commerce "transactions" as measured by estimated total e-commerce, the bottom right plot in Fig. 3.1 shows major oscillations without an obvious extrapolation into the future.

The point is that unlike domains in which we can measure progress against a stationary environment, cyber security must be viewed as an ongoing sequence of moves, countermoves, deceptions and strategic adaptations by the various actors involved—attackers, defenders, vendors and decision/policy makers. Accordingly, we believe that the appropriate science for understanding the evolving landscape of cyber security is not the logic of formal systems or new software engineering techniques. Instead, it is an emerging subarea of game theory that investigates dynamics in adversarial situations and the biases of competing human agents that drive those dynamics (see [8, 9, 23] for example).

3.1.1 Adversarial Behavior Analytics vs Classical Game Theory "Solutions"

The original goals of Game Theory were to model adversarial environments and to optimize strategies for operating in those environments. This would seem ideal for modeling cyber operations as well as other national security situations—indeed, there is a community of researchers currently investigating the application of classical Game Theory to information assurance and cyber operations.

However, the overwhelming focus of Game Theory research over the past 60 years has been on the problem of "solving" games that are defined a priori. That is, most Game Theory research to date begins by assuming a game is already defined (namely, the players, their possible moves and payoffs) and then explores properties of optimal strategies and how to compute them. Optimality is with respect to a solution criterion such as Nash Equilibrium or Pareto Optimality [5].

An obvious and growing criticism of the classical approach is that in most real world adversarial situations players do not know who the other players are, what their possible moves might be and, perhaps most importantly, what their preferred outcomes or objectives are. Put another way, none of the players actually know the complete details of the game that they are playing! A further complication is that few people outside of the Game Theory literati know what a Nash Equilibrium is, let alone how to compute one, so they typically cannot be expected to play the Nash solution.

As a result, while Game Theory can inform us about how to play chess, checkers, poker and simple illustrative examples found in most Game Theory texts, it has not been as useful in the majority of real-world adversarial situations as one might have hoped for (see [19, 20] for an interesting discussion). New directions and ideas are needed, especially in the area of cyber security.

3.1.2   Our Approach

Adversarial behavior analytics is the empirical study of players' actions in adversarial situations. The "game" in these adversarial situations is implicit and can only be understood in terms of the moves players make and how they evolve their play in response to each other's moves [12].

We have studied historical data from a variety of cyber and national security domains such as computer vulnerability databases, offensive and defensive coevolution of wormbots such as Conficker, and US border security [22, 24]. The data show that the "success rate" or other performance metrics in these different domains oscillate over time—they are not converging to any asymptote. In fact, when players are continually responding as best they can to their opponents' play, periodic and even chaotic behaviors can be exhibited [23].

Such oscillations are indicators of and intrinsic to adversarial dynamics in complex, competitive environments. In particular, each player is adapting incrementally to the observed play of his or her opponents. This can be modeled by systems of differential equations known as *replicator equations* [9, 22].

The replicator equations are typically third-degree nonlinear so that the resulting dynamics are difficult to predict analytically. However, the inverse problem of observing behaviors and estimating parameters of the replicator equations that result in those behaviors are tractable computational problems. In particular, it is possible to observe game play and strategy evolution and then make inferences about the players' motives, costs and move options.

This kind of modeling approach can explain the non-convergent dynamics we are seeing in cyber security and will help us forecast the various players' future strategies. Recognizing and harnessing the realities of such dynamic coevolution will be a key ingredient to dominating cyber operations.

3.1.3   Organization of the Paper

After this introduction, we examine in detail the documented structure of the Conficker Worm and defenses against it. This is, to our knowledge, the first quantitative attempt to extract and highlight specific adaptations in an adversarial setting. Assessment of these moves in the context of classical solution concepts such as Nash Equilibria is then discussed and is followed by an analysis of the estimated goals and motives of Conficker's developers. We then develop the notion of a Quantitative Attack Graph (QAG) and present some generic analyses of the adversarial nature of that model.

3.2 Conficker Analysis

Drawing on published sources [3, 7, 15, 16, 21], we model the interactions between Conficker (specifically its spread and update mechanisms) and its "Ecosystem" (i.e., the networked computing substrate it operates on: Microsoft, the Internet Infrastructure, the worm analysis community) as an adversarial game between two players.

3.2.1 Conficker Internal and External State Diagram

We first analyzed Conficker's internal state diagrams in terms of armoring, update and scan/infect mechanisms. The goal was to identify vulnerable points to disable Conficker (Fig. 3.2).

One area we identified was environmental mutations. Conficker A exited upon detection of a Ukrainian keyboard locale. Conficker C's well thought out innovation, its P2P module, kills Conficker C if a debugger is detected (Fig. 3.3). We also found that manipulating the Random Number Generator affects the scan/infected IP range and the Domain Generation Algorithms IP rendezvous points for potential updates (Fig. 3.4). Subversion of software/hardware encryption/hashing functionality by triggering on the public key decryption (the RSA public key is known) disables the install of new binaries (Fig. 3.5). Finally, manipulation of elapsed time/tick count (through memory writes and/or direct clock influence) affects state transitions in all mechanisms (Fig. 3.6).

Also of interest is the susceptible host population view, representing the external state transitions. Figure 3.7 shows the migration chart of the Conficker variants, while Fig. 3.8 shows the individual Conficker A/B/C/D/E host state changes. For a general discussion on subverting end systems through subsystems, see [1].

Lack of a common naming scheme for Conficker and disagreement among analysts which release constitute new versions complicate matters somewhat. For example, the third release (Microsoft's Conficker.C) is recognized as only incremental by the SRI-based Conficker Working Group (CWG), and is not recognized at all by Symantec.

Simplified State Diagram

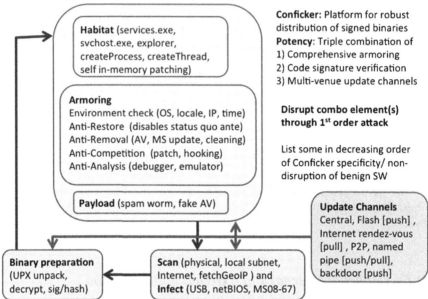

Fig. 3.2 Conficker's armoring (*green*), update (*red*) and scan/infect (*blue*) mechanisms

Simplified State Diagram

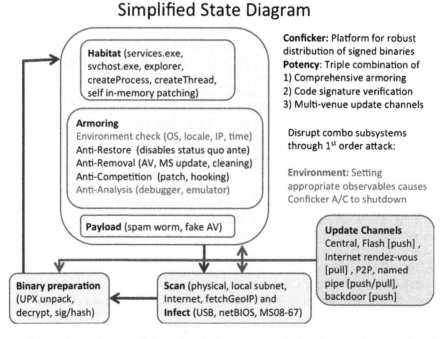

Fig. 3.3 Setting environmental observables (debugger present, VM environment, keyboard locale) causes shutdown

Simplified State Diagram
Habitat (services.exe, svchost.exe, explorer, createProcess, createThread, self in-memory patching)
Armoring
Environment check (OS, locale, IP, time)
Anti-Restore (disables status quo ante)
Anti-Removal (AV, MS update, cleaning)
Anti-Competition (patch, hooking)
Anti-Analysis (debugger, emulator)
Payload (spam worm, fake AV)
Conficker: Platform for robust distribution of signed binaries
Potency: Triple combination of
1) Comprehensive armoring
2) Code signature verification
3) Multi-venue update channels
Disrupt combo subsystems through 1st order attack:
RNG: Direct manipulation affects rendezvous points (DGA), IP scan range
Update Channels
Central, Flash [push], Internet rendez-vous [pull], P2P, named pipe [push/pull], backdoor [push]
Binary preparation (UPX unpack, decrypt, sig/hash)
Scan (physical, local subnet, Internet, fetchGeoIP) and
Infect (USB, netBIOS, MS08-67)

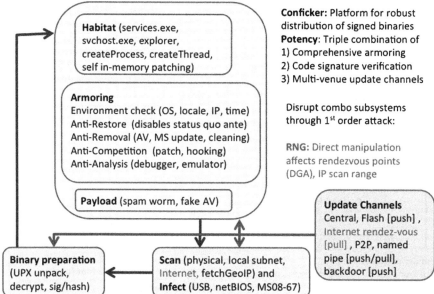

Fig. 3.4 Manipulating the Random Number Generator affects the scan/infected IP range and the Domain Generation Algorithms IP rendezvous points

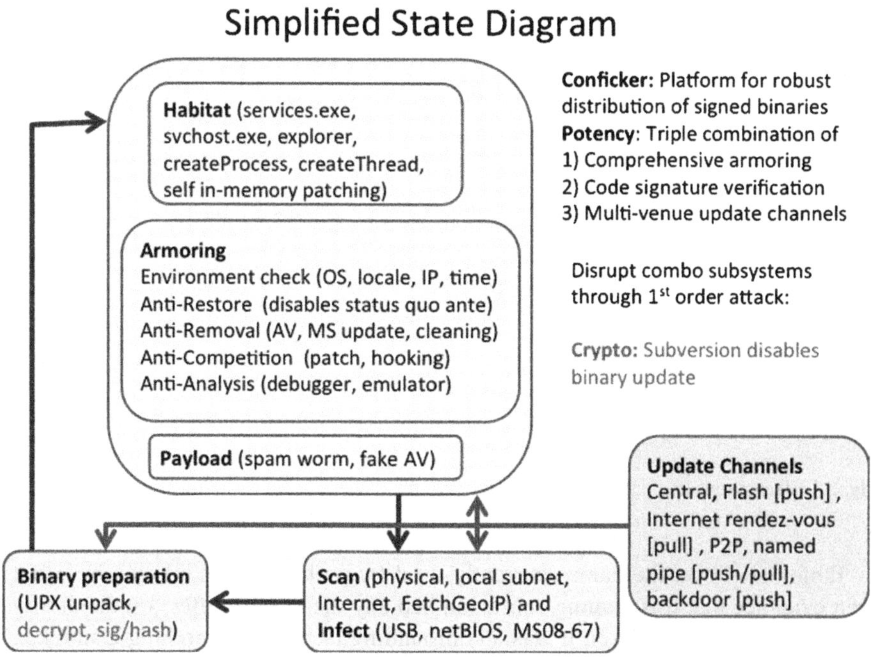

Simplified State Diagram
Habitat (services.exe, svchost.exe, explorer, createProcess, createThread, self in-memory patching)
Armoring
Environment check (OS, locale, IP, time)
Anti-Restore (disables status quo ante)
Anti-Removal (AV, MS update, cleaning)
Anti-Competition (patch, hooking)
Anti-Analysis (debugger, emulator)
Payload (spam worm, fake AV)
Conficker: Platform for robust distribution of signed binaries
Potency: Triple combination of
1) Comprehensive armoring
2) Code signature verification
3) Multi-venue update channels
Disrupt combo subsystems through 1st order attack:
Crypto: Subversion disables binary update
Update Channels
Central, Flash [push], Internet rendez-vous [pull], P2P, named pipe [push/pull], backdoor [push]
Binary preparation (UPX unpack, decrypt, sig/hash)
Scan (physical, local subnet, Internet, FetchGeoIP) and
Infect (USB, netBIOS, MS08-67)

Fig. 3.5 Subversion of encryption functionality disables the install of new binaries

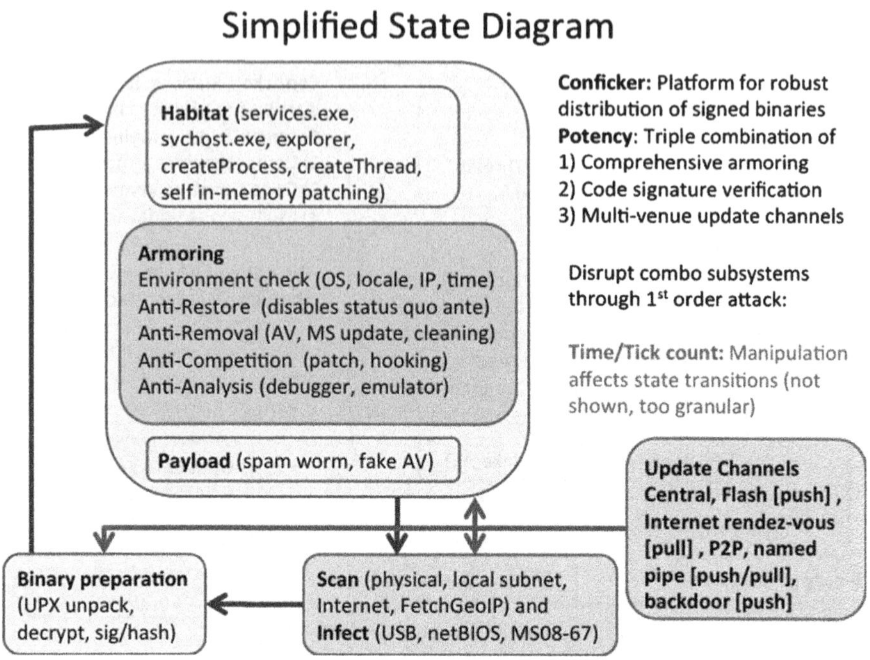

Simplified State Diagram

Habitat (services.exe, svchost.exe, explorer, createProcess, createThread, self in-memory patching)

Armoring
Environment check (OS, locale, IP, time)
Anti-Restore (disables status quo ante)
Anti-Removal (AV, MS update, cleaning)
Anti-Competition (patch, hooking)
Anti-Analysis (debugger, emulator)

Payload (spam worm, fake AV)

Binary preparation (UPX unpack, decrypt, sig/hash)

Scan (physical, local subnet, Internet, FetchGeoIP) and **Infect** (USB, netBIOS, MS08-67)

Conficker: Platform for robust distribution of signed binaries
Potency: Triple combination of
1) Comprehensive armoring
2) Code signature verification
3) Multi-venue update channels

Disrupt combo subsystems through 1st order attack:

Time/Tick count: Manipulation affects state transitions (not shown, too granular)

Update Channels
Central, Flash [push] , Internet rendez-vous [pull] , P2P, named pipe [push/pull], backdoor [push]

Fig. 3.6 Control of elapsed time/tick count affects state transitions in all mechanisms

Version Migration Chart

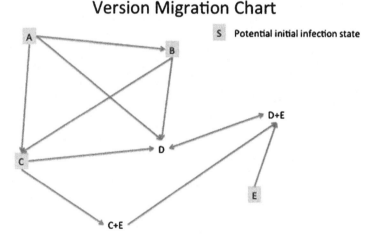

S Potential initial infection state

Fig. 3.7 Transitions and updates to Conficker variants

Table 3.1 shows the names currently used by each group. While we have not seen evidence that this naming confusion had any measurable effect on the ability to defend against Conficker, it was a confounding factor in researching Conficker. We use the Microsoft nomenclature in this report, except where noted.

Host Status A-E

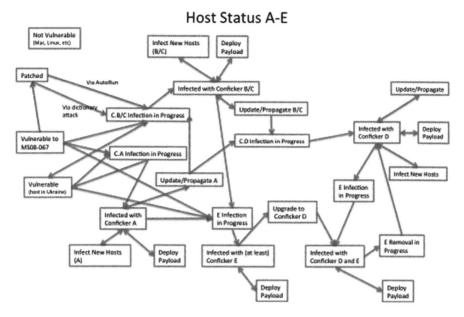

Fig. 3.8 Variant transitions and external state diagram

Table 3.1 Conficker naming conventions according to Microsoft, SRI and Symantec	Microsoft	SRI	Symantec
	Conficker.A	Conficker.A	W32.Downadup
	Conficker.B	Conficker.B	W32.Downadup.B
	Conficker.C	Conficker.B++	
	Conficker.D	Conficker.C	W32.Downadup.C
	Conficker.E	Conficker.E	W32.Downadup.E

3.2.2   Time-Evolution of Conficker

Two timelines are involved, Conficker's and the Ecosystem. Within those timelines, we distinguish among three epochs with corresponding time regimes: The era before the first appearance of Conficker A ("BCA"), followed by the emergence of the Conficker timeline ("ACA"). The final post-Conficker E ("PCE") epoch is not analyzed further in this report.

In BCA, the necessary Ecosystem pre-conditions for the viability/emergence of Conficker have to be met. These pre-conditions include "long reach propagation", a long range internet accessible vulnerability (in Conficker's case: MS08-067 RPC-DCOM), and "weaponization", an exploit for that vulnerability in a form that can be integrated into a worm (in Conficker's case: $37.80 Chinese-made exploit kit for RPC-DCOM) [17].

AS a simplifying abstraction, we view the developments in the BCA era as Ecosystem configuration fluctuations, rather than moves in a game. Once a Conficker-amenable configuration arises (i.e a worm-integratable exploit for long

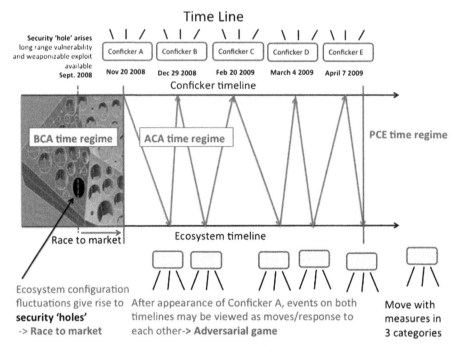

Fig. 3.9 Conficker and Ecosystem timelines. We distinguish among three epochs, but only the ACA time regime is analyzed in this paper. Race-to-market begins with an Ecosystem configuration that includes a long-range vulnerability and a concomitant worm-integratable exploit

range vulnerability), a race-to-market begins between attacker and defender to plug the security "hole". This typically takes the form of vendor software patches to fix competing with a worm exemplar to exploit the vulnerability. Events unfolding under this time regime can be modeled as a multi-objective optimization problem, balancing product performance (the "quality" of both the patch and the worm) and speed-to-market (who gets to the security hole first).

Conficker A appeared on Nov. 20, 2008. From then onwards (the ACA era), we start interpreting measures by the Ecosystem and Conficker A/B/C/D/E code evolution as moves and counter-moves in an adversarial game. We illustrate this with Fig. 3.9 below.

3.2.3 Game Moves

The game consists of sequential moves between Conficker and the Ecosystem. We fix an adversarial game with 5 rounds, consisting of 5 attacker moves and 5 responses. Moves consist of one or more measures, which are further grouped into four categories. Each Conficker move consists of the measures implemented

Player	Spread/Infect	Update	Armor
Conficker	6	12	7
Ecosystem	8	2	2

Table 3.2 Maximum pure strategy count in each category

by its respective Conficker variant (A/B/C/D/E). The measures that the Ecosystem implemented chronologically after Conficker's move are treated as the Ecosystem's response.

We group measures into Spread/Infect if they affect Conficker's spread and infection mechanisms in the wider sense. Similarly, measures that affect Conficker's update mechanism are grouped under Update, and anti-analysis/self-defense measures under Armor. Measures that do not fit into prior categories are grouped under Other.

A count of measures is given below in Table 3.2. In game-theoretic nomenclature, measures represent strategies. In each of the 5 rounds, players can select one or more strategies per category. For example, in the Spread/Infect category, Conficker E can maximally choose from the power set of 6 pure strategies (i.e $2^6 = 64$ possibilities), while the Ecosystem can maximally choose from the power set of 8 pure strategies (i.e. $2^8 = 256$ possibilities).

However, chronology excludes some possibilities: For instance, the Ecosystem cannot use strategy AVsigD as a response to Conficker A, since that anti-viral signature for Conficker D could not yet have been developed. The actual strategies chosen in each round are given in Table 3.3 (new measures are **bolded**, dropped measures are ~~struck-through~~). For example, Conficker chose three out of six pure strategies MS08-067 and EnvCheck, FetchGeoIP for Spread/Infect, and Ecosystem responded with three strategies MSpatch, AVsigA, DenyGeoIP out of 8 available pure strategies.

3.3 Nash Equilibrium or Myopic Best Response?

In Game Theory, a Nash Equilibrium arises when, in full knowledge of the other player's strategy, a player cannot benefit by changing his own strategy.[1] The notion of "benefit" is captured with payoffs for each strategy; these payoffs are in turn related to the goals of the players. To determine whether the actual strategies chosen in each game round between Conficker and the Ecosystem could be interpreted as achieving Nash Equilibria (given plausible payoff/utility functions), we examine the moves between the Ecosystem and Conficker A in the category Spread/Infect.

---

[1] This implies a Nash Equilibrium test: If, after revealing the player's strategies to one another, no player changes his strategy, despite knowing the actions of his opponents, a Nash Equilibrium has been reached.

Table 3.3 Measures implemented by Conficker and Ecosystem between November 2008 and April/May 2009

Time Period	Player	Spread/Infect	Update	Armor	Other
Nov 20, 2008–Dec 28, 2008	11/20/08 Conficker.A	**MS08-067** **EnvCheck** **FetchGeoIP**	**central** **rnd250-5** **RC4** **RSA-1024**	**obfusc**	**AVXP**
	Ecosystem response	**AVsigA** **DenyGeoIP** **MSpatch**	**blockBiz**		
Dec 29, 2008–Feb 19, 2009	12/29/08 Conficker.B	~~FetchGeoIP~~ **InclGeoIP** ~~EnvCheck~~ **LocalShare** **USB** MS08-067	~~central RC4~~ ~~RSA-1024~~ **rnd250-8** **MSbkcdr** **MD6v1** **RSA-4096**	obfusc **DNSblock** **AutoUpdDis** **AnlsShut**	
	Ecosystem response	**AVsigB** DenyGeoIP		**SRI-AB**	**MSBounty**
Feb 20, 2009–Mar 3, 2009	02/20/09 Conficker.C	LocalShare USB MS08-067	~~MD6v1~~ **MD6v2** **rnd50k-8** MSbackdoor **namedpipe**	obfusc DNSblock AutoUpdDis	
	Ecosystem response	**AVsigC** DenyGeoIP	**block250**	SRI-AB	**CWGform** MSBounty

Period	Event				
Mar 4, 2009–Apr 6, 2009	03/04/09 Conficker.D	~~LocalShare USB~~ ~~MS08-067~~	**rnd50k-110** ~~MSbackdoor~~ ~~namedpipe~~ **P2P** MD6v2	obfusc **SecServDis** **AVDis** **DNSAPI** AutoUpdDis **SRI-C** SRI-AB	
	Ecosystem response	**AVsigD** DenyGeoIP			CWGform MSBounty **WalSpyPL**
Apr 7, 2009–present	04/07/09 Conficker.E	MS08-067	MSbackdoor P2P MD6v2 Rnd50k-110	obfusc ~~SafeMdDis~~ AVDis DNSAPI AutoUpdDis SRI-C SRI-AB	
	Ecosystem response	**AVsigE** DenyGeoIP **FakeGeoIP**			CWGform MSBounty

Bolded indicates newly introduced measures. ~~Strike-through~~ indicates dropped measures

Conficker's pure strategies are EnvCheck, MS08-067, USB, LocalShare, FetchGeoIP, InclGeoIP and the Ecosystem's available pure strategies are MSpatch, AVsigA, FakeGeoIP, DenyGeoIP. Conficker implemented MS08-067, EnvCheck, FetchGeoIP and the Ecosystem responded with MSpatch, AVsigA, DenyGeoIP.

For Spread/Infect, a reasonable goal/motive for Conficker is to increase the number of infected hosts. A reasonable goal for the Ecosystem player is to reduce the vulnerable host population. A Nash Equilibrium would imply that exploiting the MS08-067 vulnerability in conjunction with the Ukraine keyboard locale check and checking IP location through the GeoIP mapping at maxmind.net gives the best payoff for Conficker in terms of increasing number of infected hosts. Conversely, issuing the MS08-067 patch, moving the web address of the GeoIP database and adding Conficker A anti-virus signature constitutes the best response to reduce the vulnerable host population. However, when Conficker A appeared in Nov/Dec 2008, a much more effective Ecosystem move to stem initial spread and contain the number of infected hosts would have been to replace the maxmind.net GeoIP lookup with a fake database (which was done months later, in June 2009). In addition, a non-realized measure such as ingress filtering of RPC TCP port 445 communications would have constituted an effective mitigation strategy.

A retrospective analysis of actual moves by Conficker and the Ecosystem suggests that they do not compute Nash Equilibria over strategy sets. We surmise this is due to the size of the strategy power sets and the incomplete information nature of the game. Instead, the players respond myopically with perceived best responses to the situation at each time step. Furthermore, that analytical framework assumes both players have the same model of the game, which may not be true.

3.3.1 Example of a Myopic Attacker Move

Conficker B introduced two new methods of self-propagation, apparently aimed at accessing networks or portions of networks not available to the randomized long range IP address vulnerability used by Conficker A. The first new method used the local area network's LocalShare, so that an infected host could in turn infect those of its local peers with insecure passwords. The second new method was to have infected hosts attempt to infect removable media such as USB keys with an attack vector via the Windows AutoRun feature.

In retrospect, the addition of the AutoRun propagation produced mixed results. As an attack vector for otherwise inaccessible systems, it proved effective: in 2009 there was not as strong a sense of security with respect to these devices, and USB drives were frequently used to bridge air gaps. However, it is not clear that bridging those air gaps was a useful thing in building a platform for releasing signed binaries, since the hosts on the other side of that gap could be infected but could not necessary access the Internet to be updated or download payloads. In Microsoft's survey of

Conficker propagation methods as a percent of attempted attacks [13], AutoRun was implicated in only 6% of attempts, suggesting that the addition of this vector was ultimately of limited utility.

Moreover, the USB drive vector was implicated in early 2009 in a string of high-profile infections such as the city of Manchester, UK and the French military. These attacks spurred greatly increased media scrutiny of Conficker, and in turn appear to have lead to accelerated adoption of anti-Conficker measures on the local level. We observed a similar increase in media interest and subsequent awareness when in 2009/2010 the Stuxnet worm spread to targets outside of Iran [11]. However, whereas Conficker's USB drive vector was in all likelihood intentionally added, it cannot be conclusively determined whether the Stuxnet leak was intentional or inadvertent.

3.3.2 Example of a Myopic Defense Move

One of the earliest moves made by the defenders in response to Conficker A was to interrupt the ability of infected hosts to retrieve the GeoIP database file from the domain maxmind.net. As part of its scanning and propagation routine, Conficker A generated a randomized list of IP addresses and checked each one against the GeoIP database file to see whether it was in Ukraine or not.

The GeoIP database file was moved by the administrators of maxmind.net to another URL shortly after the release of Conficker A. The effect of this move is difficult to gauge, as there was not an explicit study done at the time. However, experts believed that it could have slowed the propagation of the worm, and the next version of Conficker (B) included the GeoIP database along with the worm itself, even though the new location was well-known.

One of the hallmarks of the Conficker attack has been a heterogeneous botnet in which multiple older versions coexist alongside the most recent. This setup facilitated the use of a later move by Felix Leder and Tillmann Werner at the Honeynet Project to quantify the effect of this specific counterattack. Hosts infected with Conficker A that had not been upgraded to later versions were still, as of June 2009, contacting maxmind.net requesting the GeoIP file that had been taken down. Leder and Werner approached maxmind.net and had them substitute a specially prepared GeoIP file that listed the entire Internet as being in Ukraine, and then tracked the number of unique IP addresses requesting that file as an indirect way of measuring new infections.

Leder and Werner found that after the substitution (at a date in early July that they did not specify), the number of unique IP addresses contacting maxmind.net dropped precipitously, as shown in Fig. 3.10. While this move was not made during the original back-and-forth, it is suggestive that the original move may have been effective in reducing the number of new infections.

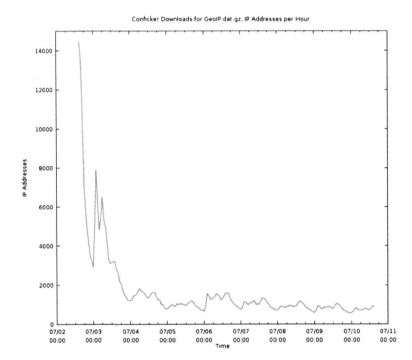

Fig. 3.10 Unique IP addresses requesting GeoIP database as a proxy for new infections after poisoning DB (Source: Honeynet Project, https://honeynet.org/node/462)

It also serves to illustrate why the original GeoIP use constituted a myopic move on the part of the Conficker creators. By using an external resource not under their control, Conficker's authors empowered the defenders with the following means:

1. An attack vector against data used for propagation;
2. A means of directly tracking new infections and thus evaluating the efficacy of successive moves.

On this occasion, this opportunity was partly missed by the defenders, who quickly found other means of attacking Conficker and who had other means of indirectly tracking its spread. However, this example is also illustrative of the kind of move that defenders need to be making, in that it not only affected the adversary, but also included a mechanism for tracking its effectiveness.

Table 3.3 outlines a shorthand list of Conficker and Ecosystem measures and a description of shorthand measures is given in Table 3.4.

We illustrate in Figs. 3.11–3.15. by means of internal state diagrams the measures that Conficker A/B/C/D/E adopted.

Table 3.4 Description of Conficker and Ecosystem measures between Nov 20, 2008 and April/-May 2009

Category	Player	Shorthand	Description
Spread/Infect	Conficker	MS08-067	Internet-accessible RPC vulnerability
		LocalShare	Local subnet Windows share drives
		USB	Local physical USB drives
		FetchGeoIP	Fetch GeoIP of IP to physical locations
		InclGeoIP	Embed GeoIP gzip file in Conficker
	Ecosystem	DenyGeoIP	Move GeoIP file to different location
		FakeGeoIP	Map all IP to Ukraine locations
		MSpatch	Software patch for MS08-067
Update	Conficker	central	Pull from trafficconverter.biz
		rnd250-5	Pull from 250 rnd domains in 5 TLDs
		rnd250-8	Pull from 250 rnd domains in 8 TLDs
		MSbckdr	Backdoor patch for MS08-067
		Rnd50k-8	Pull from 50k rnd domains in 8 TLDs
		namedpipe	Download URL transmitted to pipe
		Rnd50k-110	Pull from 50k rnd domains in 110 TLDs
		RC4	Conficker RC4 encryption
		RSA-1024	Conficker 1024 public RSA key
		RSA-4096	Conficker 4096 public RSA key
		MD6v1	first version MD6 hash implementation
		MD6v2	patched MD6 hash implementation
	Ecosystem	blockBiz	Take down trafficconverter.biz
		block250	Register all 250-5 and 250-8 domains
Armor	Conficker	DNSBlock	Block DNS lookups
		AutoUpdDis	Disable MS AutoUpdate
		SafeMdDis	Disable Windows Security Services
		AVDis	Disable AV processes
		EnvCheck	Check environmental parameters
		AnlsShut	Anti-Analysis mechanisms
		obfusc	Code obfuscation
	Ecosystem	AVsigA	Anti-virus signature for Conficker A
		AVsigB	Anti-virus signature for Conficker B
		AVsigC	Anti-virus signature for Conficker C
		AVsigD	Anti-virus signature for Conficker D
		AVsigE	Anti-virus signature for Conficker E
Other	Conficker	AVXPPL	Fake anti-virus XP payload
		CWGform	Conficker Working Group forms
		WalSpyPL	Waledac/SpyProtect payload
	Ecosystem	SRI-AB	SRI report on Conficker A/B
		SRI-C	ASRI report on Conficker A/B
		MSBounty	Microsoft $250,000 reward

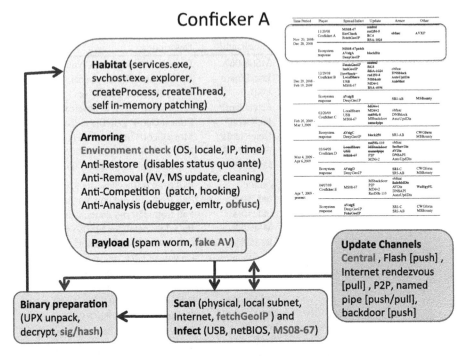

Fig. 3.11 Conficker A. Biggest Achilles heel is centralization

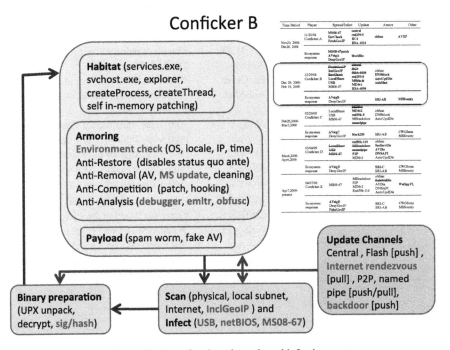

Fig. 3.12 Conficker B: Diversification of update channels and infection vectors

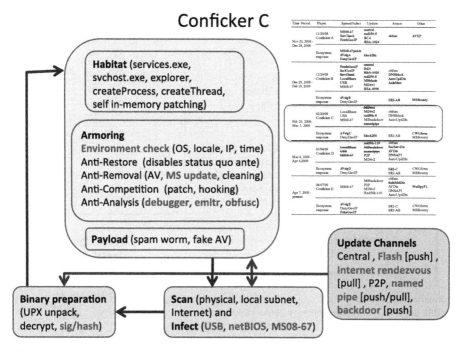

Fig. 3.13 Conficker C: More update channels, MD6 fix and extensive anti-analysis measures

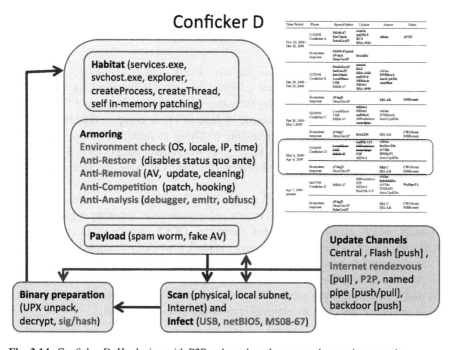

Fig. 3.14 Conficker D: Hardening with P2P, enlarged rendezvous and extensive armoring

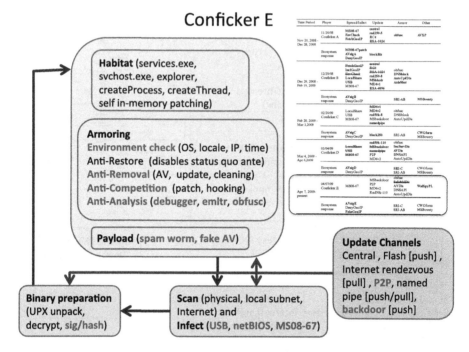

Fig. 3.15 Conficker E: Bootstrapping C to D and prototype crimeware payload

3.4 Analysis of Conficker Goals/Motives

As of February 2012, we think that the most likely goal of Conficker's creator(s) is the creation and maintenance of a large-scale reliable platform for crimeware distribution. This is corroborated by the payloads that Conficker A and E installed, as well as developments that transpired in June 2011: A group in the Ukraine was arrested for using Conficker to distribute phishing payloads to banks. The most sophisticated botnet in the world today, the 4th generation TDSS/TDL4 [6, 18], seems to be geared towards the installation of adware/spyware and spam—in other words, crimeware.

We stress that although the binary payloads so far have been crimeware, the design of Conficker enables deployment and execution of arbitrary signed binary payloads which could be used for sabotage, DDoS attacks, data destruction, intellectual property theft—virtually any payload, provided it is signed.

As a secondary plausible goal, it seems the creator(s) of Conficker wanted to preserve anonymity. Denial of attribution is the normal state of affairs for worms and malware, but it is not an absolute—some actors may not prioritize anonymity. So far, no group has credibly claimed responsibility for this worm.

3.4.1 Lessons Learned

The Lessons Learned document produced by the Conficker Working Group (CWG) essentially described the CWG's view of the final equilibrium: the botnet spread was limited, and the authors of the botnet were restricted from actually using it. We discuss three perspectives: measurement, media and goals.

3.4.1.1 Measurement

Adversarial games are played poorly in an absence of information. The ability to accurately assess the state of the game is vital to making intelligent and informed strategic decisions. This played out in Conficker domain in two ways.

The written and openly published analysis of the Conficker worm was instrumental to the defense against it. This was done mostly on an ad-hoc basis and then disseminated to blogs and security sites. This distributed effort then informed the organized effort. In particular, the identification of the randomized site generator was a crucial step in slowing the spread of Conficker.

The second type of measurement involved the ability to assess effectiveness on a large scale. The CWG concluded in their Lessons Learned report that the Conficker authors were effectively thwarted by the CWG's efforts, particularly the moves to block the update mechanism. This finding is based on the lack of a large attack and on their estimates of the size of the heterogeneous Conficker botnet over time. This estimate is made by tracking connection attempts to sinkhole IP ranges under the control of defenders.

According to this estimate, Conficker D (shown in green in Fig. 3.16) did not appear to have anywhere near the ongoing prevalence of A, B, and C combined, with a high point of over a million unique infected IP addresses identified in April 2009. Figure 3.16 shows the changes in estimates for March through December 2009, the period following the estimated release of Conficker C and the Feb 12 official launch of the CWG. Figure 3.17 shows the one year period ending February 2012 for comparison.

The CWG reports that their success on a technical level rested upon two factors: their ability to suppress the spread of Conficker D, and their ability to prevent the Conficker authors from gaining control of the A/B/C botnet. However, there is little analysis available of effective means of measuring the spread of Conficker, and there is evidence that Conficker D and E together included a masking strategy.

On a broader level, the CWG identifies their abilities to organize the defense community, get the word out about effective remediation, and to work with domain registrars on an amicable and organized basis—all crucial to their success in limiting the spread of Conficker. Although a number of the identified moves on the defense's part were unilateral (e.g. issuing the MS08-067 patch, removing the GeoIP file from maxmind.net), these organization efforts were critical to their ability to block the

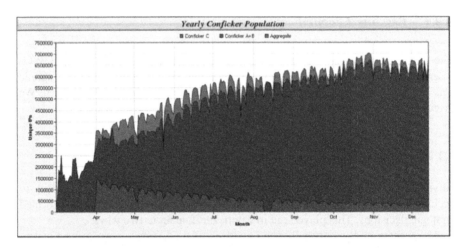

Fig. 3.16 Estimated Conficker population, Mar 2009–Dec 2009. Source: http://www. shadowserver.org/wiki/uploads/Stats/conficker-population-year.png   Copyright   ©2012   The Shadowserver Foundation. Reproduced with permission

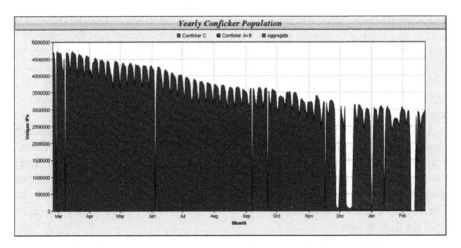

Fig. 3.17 Estimated Conficker population Mar 2011–Feb 2012. (Figure from http://www. shadowserver.org/wiki/uploads/Stats/conficker-population-year.png Copyright ©2012 The Shadowserver Foundation. Reproduced with permission

update domains. While in principle this could have eventually been done on an ad-hoc basis by the registrars themselves, in the event it took a persistent and organized team to push these changes through in a short time.

> The question naturally arises as to why remediation efforts have largely ceased. Assessments vary, but the consensus seems to be that because the botnet is not perceived to be doing anything, defenders have individually come to the conclusion that it is not worth the expenditure of resources to remove the worm from their networks (even when they know they are infected): Given any large number of infected systems, remediation becomes a very difficult task, and even harder to justify when the infection does nothing. We have no doubt that many folks infected with Conficker may not even be aware that they have been compromised. We see this issue frequently, not only with Conficker, but also with other infections that clearly do demonstrate malicious activity. Any remediation effort from the provider's perspective will be painful and lengthy. There are no easy answers here.
> *ShadowServer.net, "Conficker"*
> http://www.shadowserver.org/wiki/pmwiki.php/Stats/Conficker

This is the expected outcome, from an adversarial perspective—expending additional resources to further suppress an adversary believed to be effectively beaten makes little sense.

3.4.1.2 Media

The remediation effort against Conficker was pursued primarily by individuals and smaller organizations. Only the actual owners and operators of the hosts in question could scan their systems for Conficker and either patch uninfected hosts or download and use removal tools. However, media attention can be an effective driver of public policy decisions and in publicizing rewards. As such, the publicity surrounding Conficker is an important part of the adversarial game.

There is evidence that both sides used or manipulated media scrutiny. Microsoft's February 12, 2009 press release announcing the creation of the Conficker Working Group served to:

1. Publicize efforts, recognizing the contributions of individuals and organizations and in turn providing an incentive to continue cooperating.
2. Publicize the $250,000 reward offered by Microsoft for information resulting in the arrest and conviction of the Conficker authors.
3. Propagate links to their web site containing Conficker removal tools.
4. Attempt to spur neutral actors to work against the Conficker authors, by framing the worm as a threat to the Internet community worldwide.
5. Attempt to persuade individuals to be more vigilant in removing the worm.

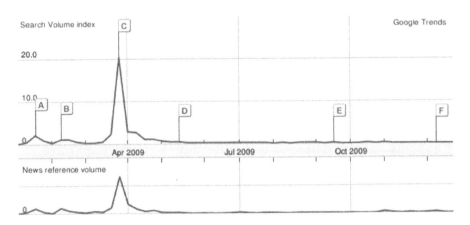

Fig. 3.18 Trend graph for search volume and news volume for "Conficker" through 2009. Data from trends.google.com

In terms of moves in an adversarial game, this was an attempt to:

1. Increase defender morale.
2. Make an attack against the persons operating the botnet.
3. Work around Conficker's DNSblock move.
4. Enlist additional resources.
5. Increase the rate at which independent actors apply remediation methods.

The effectiveness of this press release can be partly evaluated by examining the prevalence of "Conficker" as a Google search term and in news articles at the time. Labeled "B" in the Google Trends graph shown in Fig. 3.18, it resulted in a modest increase in activity. News articles on Conficker, particularly relating to a string of high-profile infections, kept the worm in the news. It is difficult to determine to what extent this affected ground-level remediation, but clearly Microsoft and other actors considered it possible.

However, the CWG did not achieve a cohesive public relations campaign. By releasing multiple press reports from individual members rather than from the working group as a whole, they largely denied themselves the ability to strategize and make deliberate moves with a minimum of leaks—a state of affairs that would have been necessary to implement, say, a comprehensive set of deception-based strategies (such as the GeoIP lookup by the HoneyNet project) [2].

There is evidence that the Conficker authors attempted to manipulate media attention. Conficker E was analyzed and determined to have a built-in event scheduled for April 1, 2009. That date is labeled "C" in Fig. 3.16, and represents a peak. After that event passed without obvious incident, there was a flurry of activity (primarily pointing out that nothing happened), and then the level of scrutiny and public awareness dropped. It is not possible to determine whether this was intentional manipulation on the part of the Conficker authors, but it is plausible as a deliberate move to reduce attention, and consistent with changes they made to the worm to give it a lower profile.

3.4.1.3 Goals

The CWG's Lessons Learned document is itself a valuable outcome—self-assessment is a vital part of determining just what game was played and what the outcome was. There is a tendency to assign goals in hindsight according to what actually was achieved; in an effort to avoid that, their Lessons Learned document contains a lengthy section giving candid feedback by members of the working group on what the goals of the group were and assessing how well they were reached.

One of the outcomes of this feedback was a frank assessment of a lack of cohesion on secondary goals such as informing the public, spearheading remediation efforts, or forming an ongoing anti-malware concern. Indeed, the very phrasing of the CWG question to its members: *"In your opinion, what were the goals of the Conficker Working Group?"* implies loose coordination. While there was consensus that a key goal of the group was "to prevent the author from updating infected computers, control of the botnet and use of it to launch a significant cyber attack", there appears to be little consensus beyond that. And while most of those interviewed agreed that their efforts had been successful in achieving this agreed-upon key goal, that assessment was not universal.

In addition, because outreach was not identified as a primary goal, the CWG was somewhat circumscribed. Local ISPs were not deliberately or formally included, and government also had almost no participation in this effort except as a recipient of information. Particularly in the case of government, there was significant internal disagreement about the proper scope of the CGW and about the role of these external actors. This disagreement served to limit the potential moves that could be made by the defense, both in terms of concerted information-gathering at the ISP level and in terms of abilities reserved to government such as intelligence-gathering and subpoena power.

3.4.1.4 Analysis of CWG's Efforts

The ability to assess the success or failure of individual actions is vital to understanding the effects of those actions and subsequently, determining an effective course of action. The CWG team's ability to observe certain IP ranges under their control, as well as their ability to coordinate certain measures amongst themselves, certainly helped.

The most successful CWG strategies were those that were clearly agreed upon: The effort to block the set of randomized domains through the individual registrars (though not perfect) worked extremely well. Conversely, goals that were harder to articulate or were not shared among all participants (such as maintaining a cohesive PR voice in order to sustain a media strategy) were not as successful.

Incorporation of corroborating data (e.g. Google Trends, web logs of infected servers) would have given a more complete picture, allowed defenders to fine-tune their actions, and may have helped avoid myopic moves. More generally, the ability to incorporate observational data from pre-established processes to link these macro-observations to adversarial moves, seems critical in hindsight.

3.5 Analytic Model of Adversarial Quantitative Attack Graphs

In this final section, we develop some analytic methods describing how attackers and defenders might reason about attack graphs. Loosely speaking, an attack graph encodes sequences of steps an attacker would need to take to achieve a desired goal against a target system. The start state of an attack graph is where an attacker begins and the end state is "goal achieved".

To illustrate the concept, consider the following notional attack graph that is actually a high level depiction of a remote attack against a computer system with the goal of exfiltrating sensitive data. Attack graphs are common constructs in computer and network security analysis [10, 14].

Attack graph analysis has traditionally only addressed reachability, not adversarial dynamics aspects involving costs and strategies by the various agents [10,14]. There have been recent attempts to quantify attack graph analysis to include costs and transition probabilities to make the technology more appropriate to risk assessment and management [4]. Our intention here is to outline an approach that begins to get at actual adversarial dynamics through attack graph analytics. To that end, consider Fig. 3.19 below, which abstracts Fig. 3.20 into states, directed transition edges and consequently paths. This attack graph has three paths whose relationships with the labeled edges are depicted Table 3.5.

The relationships between path costs, edge costs and the path–edge relationships can be expressed quantitatively using the path–edge adjacency matrix M defined as

$$M = \begin{bmatrix} 1 & 1 & 0 & 0 & 0 \\ 1 & 0 & 1 & 0 & 1 \\ 0 & 0 & 0 & 1 & 1 \end{bmatrix}$$

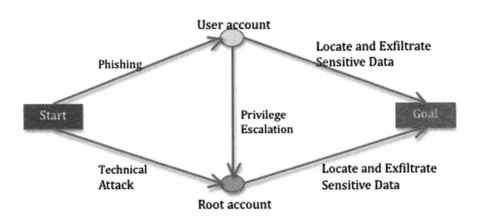

Fig. 3.19 Abstracted attack graph to illustrate the basic concepts

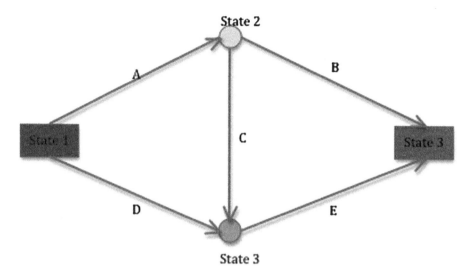

Fig. 3.20 Notional attack graph to illustrate the basic concept

Table 3.5 Path–edge relationships together with path and edge costs

	A	B	C	D	E	Path cost
Path 1	1	1	0	0	0	$Y_A + Y_B$
Path 2	1	0	1	0	1	$Y_A + Y_C + Y_E$
Path 3	0	0	0	1	1	$Y_D + Y_E$
Edge cost	Y_A	Y_B	Y_C	Y_D	Y_E	

and the edge cost matrix $Y = [Y_A\ Y_B\ Y_C\ Y_D\ Y_E]'$ where $'$ denotes transpose. The relationship is that the path costs are $M * Y$. An informed and rational attacker would choose to exploit the minimal cost path whose cost is determined by the following optimization problem and inequalities.

$$\max \alpha$$

subject to

$$M * Y \geq \alpha \mathbf{1} \tag{3.1}$$

where $\mathbf{1}$ is the column vector of all 1's. All costs are by definition non-negative.

Assume that Y is the vector of current costs the attacker has to address as given by the attack graph and the defender has a total investment of D dollars to make in protecting the system. The defender has to allocate the D units across the various edges to make the attacker's goal more costly to reach.

If we let $T = [T_A\ T_B\ T_C\ T_D\ T_E]'$ denote the allocation of resources to attack path edges and assume, for simplicity at the moment, that there is a direct linear relationship between investment and increased cost, then the defender's investment problem becomes a minimal cost attack-optimal defensive investment problem (MCA-ODI). The MCA-ODI problem is expressed as a linear program and can

be solved by standard linear programming solvers including `linprog` in the
MATLAB Optimization Toolbox.

$$\max \alpha$$

subject to

$$M * (Y + T) \geq \alpha \mathbf{1} \qquad (3.2)$$

$$T * \mathbf{1} = D \qquad (3.3)$$

such that

$$T \geq 0$$

The MCA-ODI formulation holds for every attack graph as quantified and interpreted above. Extensions to non-proportional and nonlinear relationships between
T and the minimal cost attack path can be expressed as with the generalized
formulation in Eq. (3.4):

$$\max \alpha$$

subject to

$$M * (Y + f(T)) \geq \alpha \mathbf{1} \qquad (3.4)$$

$$T * \mathbf{1} = D \qquad (3.5)$$

such that

$$T \geq 0$$

where $f(T)$ is a general nonlinear function can be solved numerically but through
more complex algorithms and with fewer analytic properties. Below, we demonstrate some simulations for the above problem with

$$Y = [Y_A \; Y_B \; Y_C \; Y_D \; Y_E]' = [5 \; 10 \; 15 \; 20 \; 25]'$$

and D varying between 0 and 100 units (Figs. 3.21–3.23).

3.6 Future Work

Future work extending the research presented in this paper includes:

1. Apply this methodology specifically to Conficker and/or other malware;
2. Use information markets or other mechanisms to quantify edge and path costs;
3. Compare and correlate computed investments with observed actions as extracted
   and analyzed in the Conficker analysis above;
4. Explore the actual functional relations between investments and attack edge cost
   increase beyond the linear relations we have used in this report;

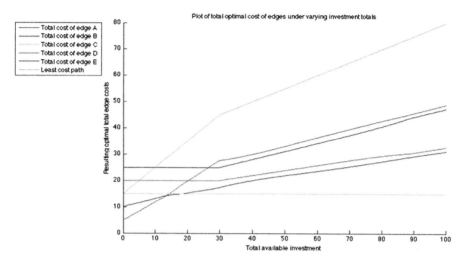

Fig. 3.21 Plot of total optimal edge costs starting with initial values specified by Y as above so the vertical axis depicts the optimal Y + T values as computed by solving the linear program. Note the changing allocations

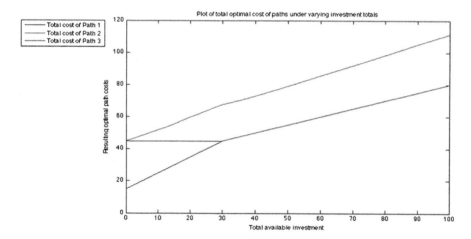

Fig. 3.22 Plot of total optimal path costs starting with initial values specified by Y as above so the vertical axis depicts the paths costs under optimal allocation as computed by solving the linear program. The minimal cost path is determined by the lowest curve

5. Relate this optimal attack graph defense investment model and solutions with previously developed network interdiction problems, max flow-min cost formulations (for scalability) and primal-dual interpretations as they might relate to Nash Equilibria or other game theoretical solution concepts;

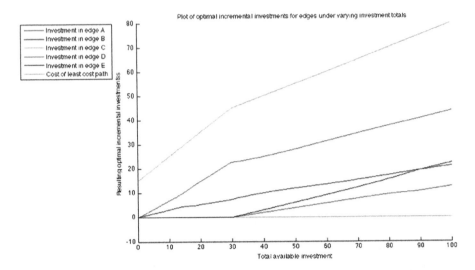

Fig. 3.23 Plot of total incremental investments per edge without the initial values as specified by Y but computed by the linear program including Y. The vertical axis depicts the edge investment allocation under optimal allocation as computed by solving the linear program. Note the different slopes and crossover point at 90 units

6. The role of deception and counter-deception in repeated games which might communicate agents' perceived notions of costs and defensive postures to mislead the adversary.

Acknowledgements We thank Vincent Berk, Patrick Sweeney, David Sicilia, Gabriel Stocco, James Thomas House and other colleagues at Process Query Systems, Dartmouth College, Siege Technologies and elsewhere for discussions and contributions that have led to these findings. This work was partially supported by DARPA Contract FA8750-11-1-0253 at Dartmouth College and US DoD contracts to Process Query Systems. All opinions and results expressed in this article are those of the authors and do not represent the positions or opinions of the US Government or sponsoring agencies.

References

1. Bilar, D.: Degradation and subversion through subsystem attacks. IEEE Security and Privacy **8**, 70–73 (2010). DOI 10.1109/MSP.2010.122. URL http://ieeexplore.ieee.org/xpl/articleDetails. jsp?arnumber=5523869
2. Bilar, D., Saltaformaggio, B.: Using a novel behavioral stimuli-response framework to defend against adversarial cyberspace participants. In: Cyber Conflict (ICCC), 2011 3rd International Conference on, pp. 169–185. CCD COE, IEEE, Tallinn, Estonia (2011). URL http://www.ccdcoe.org/publications/2011proceedings/UsingANovelBehavioralStimuli-ResponseFramework...-Bilar-Saltaformaggio.pdf
3. Bowden, M.: Worm : The First Digital World War. Grove Press (2011)

4. Carin, L., Cybenko, G., Hughes, J.: Cybersecurity strategies: The queries methodology. Computer **41**(8), 20–26 (2008). DOI 10.1109/MC.2008.295. URL http://dx.doi.org/10.1109/MC.2008.295
5. Fudenberg, D., Tirole, J.: Game Theory. The MIT Press, Cambridge MA (1991)
6. Greengard, S.: The war against botnets. Commun. ACM **55**(2), 16–18 (2012). DOI 10.1145/2076450.2076456. URL http://doi.acm.org/10.1145/2076450.2076456
7. Group, C.W.: Lessons learned. http://www.confickerworkinggroup.org/wiki/uploads/Conficker_Working_Group_Lessons_Learned_17_June_2010_final.pdf (2010)
8. Hart, S., Mas-Colell, A.: Uncoupled dynamics do not lead to Nash equilibrium. American Economic Review **93**, 1830–1836 (2003)
9. Hofbauer, J., Sigmund, K.: Evolutionary game dynamics. Bulletin (New Series) of the Amer. Math. Soc. **40**(4), 479–519 (2003)
10. Ingols, K., Lippmann, R., Piwowarski, K.: Practical attack graph generation for network defense. In: Proceedings of the 22nd Annual Computer Security Applications Conference, ACSAC '06, pp. 121–130. IEEE Computer Society, Washington, DC, USA (2006). DOI 10.1109/ACSAC.2006.39. URL http://dx.doi.org/10.1109/ACSAC.2006.39
11. Langner, R.: Stuxnet: Dissecting a cyberwarfare weapon. IEEE Security and Privacy **9**(3), 49–51 (2011). DOI 10.1109/MSP.2011.67. URL http://dx.doi.org/10.1109/MSP.2011.67
12. Levine, D.F.D.K.: The Theory of Learning in Games (Economic Learning and Social Evolution). The MIT Press, Cambridge MA (1998)
13. Microsoft: Microsoft Security Intelligence Report, Volume 12, July through December 2011. http://download.microsoft.com/download/C/9/A/C9A544AD-4150-43D3-80F7-4F1641EF910A/Microsoft_Security_Intelligence_Report_Volume_12_English.pdf (2012)
14. Phillips, C., Swiler, L.P.: A graph-based system for network-vulnerability analysis. In: Proceedings of the 1998 workshop on New security paradigms, NSPW '98, pp. 71–79. ACM, New York, NY, USA (1998). DOI 10.1145/310889.310919. URL http://doi.acm.org/10.1145/310889.310919
15. Porras, P., Saidi, H., Yegneswaran, V.: Conficker c p2p protocol and implementation. SRI International, Menlo Park, CA, Tech. Rep (2009)
16. Porras, P., Saïdi, H., Yegneswaran, V.: A foray into conficker's logic and rendezvous points. In: Proceedings of the 2nd USENIX conference on Large-scale exploits and emergent threats: botnets, spyware, worms, and more, LEET'09, pp. 7–7. USENIX Association, Berkeley, CA, USA (2009). URL http://dl.acm.org/citation.cfm?id=1855676.1855683
17. Ren, H., Ong, G.: Exploit ms-08-067 bundled in commercial malware kit. http://www.avertlabs.com/research/blog/index.php/2008/11/14/exploit-ms08-067-bundled-in-commercial-malware-kit/ (14 Nov 2008)
18. Rodionov, E., Matrosov, A.: The evolution of TDL: Conquering x64. http://go.eset.com/us/resources/white-papers/The_Evolution_of_TDL.pdf (2011)
19. Rubinstein, A.: http://arielrubinstein.tau.ac.il/papers/afterwards.pdf (2007)
20. Rubinstein, A.: Theory of Games and Economic Behavior (Commemorative Edition). John von Neumann and Oskar Morgenstern (with an introduction by Harold Kuhn and an afterword by Ariel Rubinstein). Princeton University Press, Princeotn NJ (2007)
21. Shin, S., Gu, G.: Conficker and beyond: a large-scale empirical study. In: Proceedings of the 26th Annual Computer Security Applications Conference, ACSAC '10, pp. 151–160. ACM, New York, NY, USA (2010). DOI 10.1145/1920261.1920285. URL http://doi.acm.org/10.1145/1920261.1920285
22. Sicilia, D., Cybenko, G.: Application of the replicator equation to decision-making processes in border security. Proceedings of SPIE Defense and Security, 2012, Baltimore MD (2012)
23. Sparrow, C., van Strien, S., Harris, C.: Fictitious play in 3 x 3 games: the transition between periodic and chaotic bahavior. Games and Economic Behavior **63**, 259–291 (2008)
24. Sweeney, P., Cybenko, G.: An analytic approach to cyber adversarial dynamics. Proceedings of SPIE Defense and Security, 2012, Baltimore MD (2012)

Chapter 4
From Individual Decisions from Experience to Behavioral Game Theory: Lessons for Cybersecurity

Cleotilde Gonzalez

Abstract This chapter discusses a central challenge arising from the success of modeling human behavior in making decisions from experience: our ability to scale these models up to explain non-cooperative team behavior in a dynamic cyber space. Computational models of human behavior based on the Instance-Based Learning Theory (IBLT) have been highly successful in representing and predicting individual's behavior of decisions from experience. Recently, these IBL models have been also applied to represent a security analyst's experience and cognitive characteristics that would result in accurate predictions of threat identification and cyber-attack detection. The IBL models derive predictions on the accuracy and timing of threat detection in a computer network (i.e., cyber situation awareness or cyberSA). This chapter summarizes the current state of models at the individual level, and it describes the challenges and potentials for extending them to address predictions in 2-player (i.e., defender and attacker) non-cooperative dynamic cybersecurity situations. The advancements that would potentially contribute to a more secure cyberspace are discussed.

4.1 Introduction

Cybersecurity is a non-cooperative "game" that involves strategic interactions between two "players": the defender of the network (e.g., cyber security analyst) and the attacker (e.g., hacker). This is non-cooperative because it is assumed that the defender and the attacker act independently, without collaboration or communication with each other. Recent research on adversarial reasoning is now being used and

C. Gonzalez (✉)
Dynamic Decision Making Laboratory, Department of Social and Decision Sciences
Carnegie Mellon University, Pittsburgh, PA 15213, USA
e-mail: coty@cmu.edu

S. Jajodia et al. (eds.), *Moving Target Defense II: Application of Game Theory and Adversarial Modeling*, Advances in Information Security 100, DOI 10.1007/978-1-4614-5416-8_4, © Springer Science+Business Media New York 2013

applied to many new and exciting domains, including counter-terrorism, homeland security, health and sustainability, and more recently to cybersecurity [46].

In cybersecurity research, many technologies based on Economic Game Theory have become available to support the cybersecurity analyst. For example, a recent survey of Game Theory applied to cybersecurity [44] summarizes interesting technical solutions designed to enhance network security. However, the survey concludes that the current game-theoretic approaches to cybersecurity are based on either static game models or games with perfect or complete information; thus, misrepresenting the reality of the network security context where situations are highly dynamic and where security analysts must rely on imperfect and incomplete information. Although more technologies have become available, these are not implemented to account for the social and cognitive factors that determine the job of the cyber analyst (and of the attacker). These tools are often incomplete and ineffective to support the analysts job and to predict accurately the attacker's strategies. In fact, analysts often require of a multitude of tools and technologies that may even result in higher cognitive workload on the job. Thus, the current technology often ignores the analyst's social and cognitive limitations; their understanding of the hacker's strategy and projections of possible actions; and the dynamics of an evolving strategic situation (i.e., they lack Cyber Situation Awareness, Cyber-SA).

We aim at addressing the current shortcomings of cybersecurity research by capitalizing on the success of models of individual decision making in dynamic and uncertain situations. These models, based on the Instance-Based Learning Theory (IBLT) [21], have demonstrated accurate and robust explanations and predictions of human behavior in multiple contexts including dynamically complex tasks [20, 36]; training paradigms of simple and complex tasks [17–19]; simple stimulus-response practice and skill acquisition tasks [11]; repeated binary-choice tasks [18, 28, 30]; and more recently, cybersecurity [7, 8, 10]. The main advantages of IBL models are their accurate representation of human information processing in dynamic and interactive environments [18].

In this chapter, the foundations of IBLT and models of decision from experience are summarized. Then, current search involving predictions from IBL models in the cybersecurity context are discussed. This chapter then turns to the important current challenges involving Behavioral Game Theory in cybersecurity, and how to scale IBL models of individual decision from experience up to representing human behavior in multi-player non-cooperative games. Finally, the chapter concludes with a discussion of how the proposed escalation of the IBL models is expected to contribute to a more secure cyberspace.

4.2   Instance-Based Learning Theory and Models of Decisions from Experience

IBLT was developed as a descriptive theory to explain and predict learning and decision making in real-time dynamic tasks [21]. Dynamic tasks are characterized by decision conditions that change spontaneously and as a result of previous decisions while attempting to maximize gains over the long run [12, 39]. Dynamic tasks range in their levels of dynamic characteristics [12]: from the least dynamic task that involves sequential but independent decisions in an environment where neither the environment nor the participants information about it is affected by their previous decisions; to the most dynamic task in which sequential decisions are made while an environment and the person's information about it changes over time and as a function of previous decisions [12].

IBLT [21] uses a cognitive information representation called an instance. An instance consists of three parts: a situation (a set of attributes that define the decision situation), a decision or action, and an outcome or utility of the decision in the situation. The different parts of an instance are built through a general decision process including: characterizing a situation by attributes in the task, making a decision based on expected utility, and updating the utility in the feedback stage. The instances accumulated over time in memory are retrieved and used repeatedly in future decisions. Their strength in memory, called activation, is reinforced according to statistical procedures. These statistical procedures were originally developed by Lebiere [27], and Anderson and Lebiere [2] as part of the ACT-R cognitive architecture.

In making a choice, IBLT recommends the selection of the alternative with the highest *blended value*, V [17, 30], which is computed from all past instances where the outcome of that alternative was observed. The blended value of alternative j is defined as:

$$V_j = \sum_{i=1}^{n} p_i x_i \qquad (4.1)$$

where x_i is the value of the observed outcome of alternative j in instance i, and p_i is the probability of retrieving instance i from memory. The blended value of an alternative is thus the sum of all observed outcomes x_i of corresponding instances in memory, weighted by their probability of their retrieval from memory. In any trial, the probability of retrieving instance i from memory is a function of that instance's activation relative to the activation of all other instances corresponding to that alternative, given by

$$p_i = \frac{e^{A_i/\tau}}{\sum_j e^{A_j/\tau}} \qquad (4.2)$$

where τ is random noise defined as $= \sigma 2$, and σ is a free noise parameter. Noise in Eq. (4.2) captures the imprecision of recalling instances from memory.

The activation of each instance in memory depends upon the *Activation* mechanism from ACT-R [2]:

$$A_i = ln\left(\sum_{j=1}^{n} t_{ji}^{-d}\right) + \sum_{j=1}^{m} W_j.(S - ln(fan_{ji})) - \sum_{l=1}^{k} P_l.M_{li} + N_i(0,\sigma) \qquad (4.3)$$

The activation A_i of an instance i reflects the availability of an instance in memory. The activation is determined by the base-level activation $ln(\sum_{j=1}^{n} t_{ji}^{-d})$, the spreading activation $\sum_{j=1}^{m} W_j.(S - ln(fan_{ji}))$, the partial matching $\sum_{l=1}^{k} P_l.M_{li}$, and noise $N_i(0, \sigma)$.

The base-level activation of an instance i reflects its recency and frequency of use in past trials/times. n is the total number of previous uses of that instance (frequency), including when the instance was first created, t_{ji} is the time elapsed since the jth use of the instance i (recency), and d is the memory decay rate (a free parameter with a default value of 0.5). A high value of d (e.g., $d > 2.0$) means rapid decay of memory (forgetting easily) and reliance on recent past experiences, whereas a low value of d means slow decay of memory and reliance on temporally distant experiences. An IBL model that uses only this base-level activation has been highly successful in accounting for and predicting behavior in dual-choice tasks under different paradigms of decisions from experience [17, 18, 30].

The spreading activation reflects the impact of a task situation's attribute j on the instance i in memory. m is the total number of attributes in the situation, and W_j represents the amount of activation from attribute j to instance i. ACT-R assumes there is a limited total amount of attention W (a free parameter, with value of 1.0 as default), which gets divided equally among all the m attributes of the situation (thus, the W parameter is the sum of all W_j) [1,3]. Thus, the W parameter influences the attention to relevant and irrelevant attributes in the task situation [3, 32]. The term $(S - ln(fan_{ji}))$ is called the strength of association. S (a free parameter that can be set to any value, and whose default value of "nil" in ACT-R means not using spreading activation) is the maximum associative strength, and fan_{ji} is the number of instances in memory in which the attribute j appears as part of the instance attributes plus one (as attribute j is associated with itself in instance i).

The partial matching component represents the mismatch between the attributes of a situation in the task and the attributes of the instance i in memory. k is the total number of attributes for a task situation that are used to retrieve instance i from memory (thus, the summation is only computed over those attributes of an instance that correspond to the task situation attributes used to retrieve the instance from memory). The match scale (P_l) reflects the amount of weighting given to the similarity between attribute l of an instance i in memory and the corresponding situation attribute in a task. P_l is generally defined to be a negative integer with a common value of -1.0 for all attributes k of an instance i (but it could also be a free parameter with different negative real values for different attributes of an instance i). The M_{li} term, or match similarity, represents the similarity between the value l of an attribute in a task situation (i.e., the situation attribute l from the task that is used to retrieve the instance i from memory) and the value in the corresponding attributes of the instance i in memory. Typically, M_{li} is defined using a linear similarity function that assumes M_{li} to be equal to the absolute value of the difference between a situation attribute from the task and the corresponding attribute value of an instance. Thus, the $\sum_{l=1}^{k} P_l.M_{li}$ specification defines the similarity between the task current situation attributes to the attributes of potentially retrievable instances from memory. Finally, the noise component is a Gaussian function with a mean of 0 and a standard deviation of σ.

IBLT and the IBL model that represents the human memory mechanisms described above have been applied to a variety of domains. Recently we have applied the IBL models to the cybersecurity context [7, 8, 10]. Next, this research in progress is summarized.

4.3   IBL Models in the Cybersecurity Context

IBL models in the cybersecurity context have been used to represent the cognitive processes involved in the work of a cybersecurity analyst, a decision maker in charge of protecting a corporate network (e.g., an online retail company with an external webserver and an internal fileserver) from threats of random or organized cyber-attacks. During a cyber attack, there could be both malicious network events (threats) and benign network events (non-threats) occurring in a sequence. Threats are generated by attackers, while non-threats are generated by friendly users of the network. Our IBL models have focused on representing the recognition and comprehension of threats while monitoring a set of network events. In order to accurately and timely detect cyber attacks, an analyst often relies on highly sophisticated technologies that aid her in the detection of threats [24]. The general idea is that IBL models may be used to obtain predictions regarding the accuracy and timeliness of threat detection, which are two measures of cyber situation awareness (CyberSA) [8]. CyberSA involves situation recognition, the perception of the type of cyber attack, source (who, what) of the attack, and target of the attack; situation comprehension, understanding why and how the current situation is caused and what is its impact; and situation projection, determining the expectations of a future attack, its location, and its impact [24, 29].

The general technical approach we follow involves the comparison of IBL model's predictions in a particular task to the human behavior in the same task. Figure 4.1 presents this approach. In this case, a threat detection task is performed by a human participant in behavioral studies, and data collected from these studies are compared to the predictions made by a cognitive model created according to IBLT. The cognitive model represents the process by which a human analyst makes decisions regarding a cyber attack, after perceiving a sequence of possible threat and non-threat events. The model performs the same task as the human participant. Data resulting from the human and model interactions with the task are then compared to each other at different levels of detail. This comparison is informative and result in improvements to the cognitive representations as well as in new ideas for behavioral experiments.

A challenge to following this technical approach in the context of cybersecurity is the reduced number of security analysts that we can use for model comparison: the high demand for analysts, their lack of availability for laboratory experiments, and

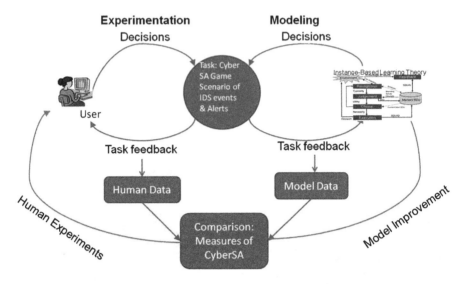

Fig. 4.1 Cognitive modeling and laboratory experimentation technical approach

the difficulty of studying real-world cyber-security events. To address this challenge we have followed two routes: (1) produce only predictions from computational models that could be verified or used as guidelines for training and selection of real-world analyses, and (2) reduce the complexity of the cybersecurity tasks to be able to obtain human data from novices that can be compared to model predictions. We have made some progress on both fronts, but much of the human data collection is still ongoing. Although not yet validated, however, the IBL model has been used to make interesting predictions regarding cybersecurity human behavior, particularly representing a cybersecurity analyst. The model and current predictions are summarized next.

4.4 IBL Model Predictions of Cybersecurity Behavior

We have built IBL models of human behavior representing the influence of three human factors in a cybersecurity task [8, 10]: The mix of threat and non-threat experiences stored in memory; the analyst tolerance to threats (i.e., how many network events an analyst perceives as threats before deciding that these events represent a cyber attack); and adversarial behavior (i.e., an attacker strategy).

Prior research indicates that the analyst cyberSA is likely a function of experience with cyber attacks [7, 24]. For example, Dutt et al. [7] and Dutt and Gonzalez [10] have provided initial predictions about the cyberSA of a simulated analyst according to its experience and tolerance. Dutt and Gonzalez [10] created a cognitive model of an analyst's cyberSA based upon IBLT and populated the model memory with threat and non-threat experiences. The model tolerance was determined by the number of events perceived as threats before it declared the sequence of network events to be a cyber attack. Accordingly, a model with a greater proportion of threat experiences is likely to be more accurate and timely in detecting threats compared to one with fewer experiences. That is because, according to IBLT, possessing recent and frequent past experiences of network threats also increases the model opportunity to remember and recall these threats with ease in novel situations.

Prior research has also predicted that an analyst tolerance to threats is likely to influence her cyberSA [10, 37, 45]. For example, Salter et al. [45] highlight the importance of understanding both the attacker and the analyst's tolerance, and according to Dutt and Gonzalez [10], an analyst is likely to be more accurate when her tolerance is low rather than high. That is because possessing a low tolerance is likely to cause the analyst to declare cyber attacks very early on, which may make an analyst more timely and accurate in situations actually involving early threats. Although the studies discussed provide interesting predictions about a simulated analyst experience and tolerance, they do not consider the role of adversarial behavior (i.e., attacker strategies). These studies also do not consider how the analyst behavior might interact with the adversary behavior to influence the analyst cyberSA. The influence of adversarial behavior may be very different from the influence of the analyst's memory factors. First, actions from the attacker are external or outside of the analyst's control. Second, previously encountered adversarial behaviors might influence analyst current experiences and tolerance to threats. Adversarial behavior can be accounted for through the representation of different strategies. Depending on these strategies, threats within a network might occur at different times and their timing is likely to be uncertain [24]. For example, an impatient attacker could execute all threats very early on in a cyber attack; whereas, a patient attacker might decide to delay the attack and thus threats would appear late in the sequence of network events. Given the influence of information recency on people decisions [7, 9, 21], an analyst with a greater proportion of threat experiences and a low tolerance is expected to be more accurate and timely compared to an analyst with fewer threat experiences and a high tolerance. However, we do not expect that to be the case for a patient attack strategy. That is because according to IBLT, a model (representing an analyst) will make detection decisions by recalling similar experiences from memory. When the model has more threat experiences in memory, it is more likely to recall these experiences early on, making it accurate if threats occur early in an attack (i.e., generated by an impatient strategy). The activated threat experiences would be recalled faster from memory and are also likely to increase the chances that the accumulation of evidence for threats exceeds the model low tolerance quickly. By the same logic when threats occur late in cyber attacks (i.e., generated by a patient strategy), the situation becomes detrimental to the accuracy and timeliness of a model that possesses many threat experiences in memory and has a low tolerance. In summary, a model experience of threats, its tolerance to threats, and an attack strategy may limit or enhance the model cyberSA.

Our current findings [10] indicate that the model accuracy (d and scenario hit rate) was greater when the attacking strategy was impatient rather than patient; however, the model needed a greater proportion of attack steps against the impatient attack strategy. Also, the model predicted accuracy (scenario hit rate and d) was greater when it was risk-averse rather than risk-seeking. For a risk-averse model, the accumulation of threat evidence takes less time to build, and thus it stops earlier in the sequence compared with a risk-seeking model (in support of this explanation, the proportion of attack steps in our results was generally low for a model that was risk-averse). Therefore, it is likely that observing a fewer number of network events, due to being risk-averse, also reduces the model chance of committing more false-alarms relative to hits, and this observation increases the accuracy. There was also an interaction between different attack strategies that differed in the timing of threats and the model experiences: For an impatient attack strategy, possessing threat-prone experiences helped the model accuracy (due to high hit rates); whereas, for a patient attack strategy, possessing threat-prone experiences hurt its accuracy (due to high false-alarm rates). Again, this result is expected given that when threats occur early on, possessing a majority of threat instances increases the likelihood of detecting these threats early on. Moreover, based upon the same reasoning, increasing the likelihood of detecting threats causes the model to detect these threats earlier, which hurts the accuracy when these threats actually occur late in the attack.

Although intriguing, the predictions cannot be taken as granted, and they need to be corroborated with human data. Current efforts are directed towards these next steps. The next section describes how these models of individual behavior can be extended to account for situations described in Behavioral Game Theory in the context of cybersecurity.

4.5   Behavioral Game Theory and Cybersecurity

Many models, in addition to IBL models, have explained individual behavior in decisions from experience. Reinforcement models of learning are perhaps the most common in the literature. These models often assume that choices are reinforced based on immediate feedback [15, 43]. Most of the current models of individual learning and decision making behavior emphasize a weighted adjustment process of the outcomes (mostly economic outcomes), by which the value of a previously observed outcome is combined or updated with a newly observed outcome [5, 13, 23, 34]. In contrast to a purely economic perspective, a Behavioral Game Theory (BGT) perspective explains how conflict and cooperation are impacted by social dimensions such as interpersonal relationships, past encounters, identities, and emotions [42, 47, 48]; and it is important to disentangle the influence of the above social concerns and to go beyond monetary outcomes for individuals [6]. Drawing upon experience in the real-world, people may implicitly care about an ongoing relationship and the image they project to others, despite being told to disregard these factors. That is, people make decisions from experience in social interactions, and feedback may reinforce some social outcomes and behaviors over others. For example, real-world social dilemmas are often described as abstract games where an individual's goal is to maximize economic benefit by cooperating or competing with others. In the well-known Prisoner's Dilemma (PD) [4, 40], each of two players chooses to "cooperate" or "defect" given information about the payoffs if one or the other action are chosen by each of the players. In the Iterated Prisoner Dilemma (IPD) where players interact repeatedly, computing the best response to maximize total payoffs becomes more complex as additional strategic factors come into play, including the possibility to punish defection or reciprocate cooperation [38]. Deviations from the mutual defection equilibrium in the one-shot PD could result from inherent altruism, as well as from norms prescribing fairness and kindness. Just as economic considerations become more complex in an IPD played for multiple rounds, so too do the social considerations become more complex in repeated interaction.

One important social aspect of conflict relates to the information available to decision makers, which ranges from no information about interdependence with others to complete information about the actions of others, their influence on the other's outcomes, and the cultural identities of others [22, 35]. These informational characteristics influence reciprocation, fairness, trust, power, and other social concerns. Not surprisingly, the effects of these social factors on individual and collective behavior are difficult to capture in computational models that have oftentimes been designed to account primarily for individual decisions and outcomes.

In cybersecurity situations, the social information that analysts have about the attackers can vary dramatically, and thus, influence both behavior and joint outcomes. Some conflict situations occur with minimal information, as in cases where individuals affect one another outcomes without even realizing their interdependence. At the opposite end of the information spectrum, some conflict situations occur with substantial levels of information when individuals know one another identities and motivations. The cybersecurity context is quite uncertain and little information exists regarding the attacker intentions and strategies.

Many models of individual decisions from experience are incapable of representing response to social contexts. For example, Erev and Roth [16] noted that simple reinforcement learning models predicted the effect of experience in two-person games like the IPD only in situations where players could not punish or reciprocate. A simple model predicts a decrease in cooperation over time even though most behavioral experiments demonstrate an increase in mutual cooperation due to the possibility of reciprocation [40, 41]. To account for the effects of reciprocation, Erev and Roth made two explicit modifications to the basic reinforcement learning model: if a player adopts a reciprocation strategy, he will cooperate in the next trial only if the other player has cooperated in the current trial; the probability that a player continues with the same strategy will depend on the number of times the reciprocation strategy was played [16]. Although these tweaks to the model may accurately represent the kind of cognitive reasoning that people actually use in the IPD, they are unlikely to generalize to other situations with different action sets or outcomes.

Unlike models that require explicit modification to predict human behavior in conflict situations, we have started to use IBL models that offer the potential for generalization even in their standard form, without predefined implementations of interaction strategies [18]. Although most of the tasks modeled to IBLT are individual tasks, there have been some initiatives to use IBLT in multi-person games. For example, Gonzalez and Lebiere [20] reported a cognitive model of an IPD, initially reported by Lebiere, Wallach, and West [29], that assumes instances are stored in memory, including one own action, the other player action, and the payoff. The decision is made according to a set of rules that, given each possible action, retrieves the most likely outcome from memory and selects the action with the highest payoff. This model was shown to accurately capture human behavior in the IPD. More recently, IBLT was also used as the basis for developing computational models that predict human behavior in more complex multi-person tasks, including a market entry game [18]. This model shares some basic features with IBL models of individual choice [31]: weighting of prior outcomes by their probability of retrieval from memory (i.e., the blending mechanism); dependence on the recency and frequency of past outcomes; and an inertia mechanism that depends upon surprise as a function of blended outcomes. In presenting a hierarchy of social information that would influence social decisions from experience, we propose some ways in which IBLT can account for more nuanced social aspects of interaction such as reciprocation, fairness, reputation, trust, and power [26].

Our current research addresses a central challenge from modeling individual decisions from experience: scaling these models up to represent the response of decision makers to social information. Our results [8] suggest that the nature of adversarial behavior in an analyst work environment would influences cyberSA. One characteristic of adversarial behavior is the attacker strategy regarding the timing of threats during a cyber attack. but regardless of the strategy there is prevailing uncertainty in terms of exactly when threats might appear in a cyber attack. Thus, it is important for the analyst to develop timely and accurate threat perception, and both the nature of the analyst and adversary behaviors may greatly influence the analyst cyberSA.

We propose a basic research program that involves studying motivational factors (e.g., costs and benefits of actions from the attacker and defender viewpoint), environmental factors (e.g., information available to players about each other), and the influence of these factors on technology constraints (e.g., how network responds based upon the defender last actions and network accuracy about reporting attacks). This research program will account for cognitive limitations that previous technical solutions have ignored, such as the defenders and attackers memory and recall limitations, and their experiential decisions. In other words, we will use a Behavioral Game Theoretical (BGT) approach to study cyber security. This research will help define how decisions from experience through repeated interactions between human players would result in the evolution of defense and attack strategies during the course of a cyber attack, and across multiple attacks and attack patterns. This study of decisions from experience helps to understand human behavior in natural and dynamic environments where one cannot solely rely on the rationality assumption or

Nash equilibriums due to social complexity [14]. In this regard, application of IBLT to the defender and attacker experiential decisions in non-cooperative games will help explain how these decisions are impacted by motivational and environmental factors and technology constraints, and will help improve current technical solutions to provide better decision support to defenders.

Our threat model is defender-centric, and it is based on the notion that any network has assets of value worth protecting. These assets have certain vulnerabilities; internal or external threats exploit these vulnerabilities in order to cause damage to the assets, and appropriate security countermeasures exist that might mitigate these threats [33]. Our threat model considers the applications of BGT to study decisions from experience in cybersecurity. We propose using simple games that are played between human players and that generalize to different scenarios that include stealing information, denial-of-service (DoS) through buffer overflow, defacing website, SYN, Teardrop, Smurf, and other virus attacks [24]. Given the lack of consideration about cognitive limitations in existing cybersecurity research, our research studies the role of motivational and environmental factors and their interaction with technology constraints on attackers and defenders decisions.

4.6 Conclusions

Describing and predicting social behavior computationally is a current research challenge with many unanswered questions: How can we account for non-monetary incentives such as fairness, trust, reciprocation, and power? Do social effects need to be explicitly represented in models that correctly predict behavior? How do individual cognitive constraints influence the use of social information? How do we scale our behavioral measures and models from the individual level to the social level? Our research program might offer a fruitful approach to modeling empirical differences in conflicts across different levels of social information.

Cyber attacks are becoming widespread [25], and it is likely that these attacks succeed due to different incentives available to attackers and defenders (motivational factors), the information availability about opponents (environmental factors), and the inefficiencies of the current technology and networks response mechanisms (technology constraints). Therefore, there is an urgent need to evaluate the effects of these factors on the resulting attack-and-defend behaviors. This knowledge will help to develop policy interventions to reduce cyber attacks and promote appropriate defense strategies. This research will provide information on actual human behavior in making decisions from experience in competitive games, where players take the role of defender or attacker, which will inform the development of algorithms and technologies that are realistic and that support the work of the main decision makers (analysts) in the cyber world. Thus, unlike the existing cybersecurity research, we will determine human behavior in these situations, while considering the cognitive limitations that are present among both human attackers and defenders.

Acknowledgements This research was a part of a Multidisciplinary University Research Initiative Award on Cyber Situation Awareness (MURI; #W911NF-09-1-0525) from Army Research Office to Cleotilde Gonzalez. We thank Hau-yu Wong for editing this manuscript.

References

1. Anderson, J. R., Bothell, D., Byrne, M. D., Douglass, S., Lebiere, C., & Qin, Y. (2004). An integrated theory of the mind. *Psychological Review, 111*(4), 1036–1060.
2. Anderson, J. R., & Lebiere, C. (1998). *The atomic components of thought.* Hillsdale, NJ: Lawrence Erlbaum Associates.
3. Anderson, J. R., Reder, L. M., & Lebiere, C. (1996). Working memory: Activation limitations on retrieval. *Cognitive Psychology, 30*(3), 221–256.
4. Axelrod, R. (1980). Effective choice in the Prisoners Dilemma. *Journal of Conflict Resolution, 24*(1), 3–25.
5. Bush, R. R., & Mosteller, F. (1955). *Stochastic models for learning.* Oxford, UK: John Wiley & Sons, Inc.
6. Colman, A. M. (2003). Cooperation, psychological game theory, and limitations of rationality in social interaction. *Behavioral and Brain Sciences, 26,* 139–198.
7. Dutt, V., Ahn, Y.-S., & Gonzalez, C. (2011). Cyber situation awareness: Modeling the security analyst in a cyber-attack scenario through instance-based learning. In Y. Li (Ed.), *Lecture Notes in Computer Science* (Vol. 6818, pp. 281–293). Heidelberg: Springer Berlin.
8. Dutt, V., Ahn, Y. S., & Gonzalez, C. (2012). Cyber situation awareness: Modeling detection of cyber attacks with Instance-Based Learning Theory. Unpublished manuscript under review.
9. Dutt, V., Cassenti, D. N., & Gonzalez, C. (2011). Modeling a robotic operator manager in a tactical battlefield. In *Proceedings of the CogSIMA 2011: 2011 IEEE International Multi-Disciplinary Conference on Cognitive Methods in Situation Awareness and Decision Support* (pp. 82–87). Miami Beach, FL: IEEE.
10. Dutt, V., & Gonzalez, C. (2011). Cyber situation awareness through Instance-Based Learning: Modeling the security analyst in a cyber-attack scenario. In C. Onwubiko & T. Owens (Eds.), *Situation awareness in computer network defense: Principles, methods and applications.* Hershey, PA: IGI Global.
11. Dutt, V., Yamaguchi, M., Gonzalez, C., & Proctor, R. W. (2009). An instance-based learning model of stimulus-response compatibility effects in mixed location-relevant and location-irrelevant tasks. In A. Howes, D. Peebles & R. Cooper (Eds.), *Proceedings of the 9th International Conference on Cognitive Modeling* - ICCM2009. Manchester, UK.
12. Edwards, W. (1962). Dynamic decision theory and probabilistic information processing. *Human Factors, 4,* 59–73.
13. Erev, I., Glozman, I., & Hertwig, R. (2008). What impacts the impact of rare events. *Journal of Risk and Uncertainty, 36*(2), 153–177.
14. Erev, I., & Haruvy, E. (in press). Learning and the economics of small decisions. In J. H. Kagel & A. E. Roth (Eds.), *The Handbook of Experimental Economics.* Princeton, NJ: Princeton University Press.
15. Erev, I., & Roth, A. E. (1998). Predicting how people play games: Reinforcement learning in experimental games with unique, mixed strategy equilibria. *The American Economic Review, 88*(4), 848–881.
16. Erev, I., & Roth, A. E. (2001). Simple reinforcement learning models and reciprocation in the Prisoner's Dilemma game. In G. Gigerenzer & R. Selten (Eds.), *Bounded rationality: The adaptive toolbox* (pp. 215–231). Cambridge, MA: MIT Press.
17. Gonzalez, C., & Dutt, V. (2011). Instance-based learning: Integrating decisions from experience in sampling and repeated choice paradigms. *Psychological Review, 118*(4), 523–551.

18. Gonzalez, C., Dutt, V., & Lejarraga, T. (2011). A loser can be a winner: Comparison of two instance-based learning models in a market entry competition. *Games, 2*(1), 136–162.
19. Gonzalez, C., Dutt, V. & Martin, J. (2011). Scaling up Instance-Based Learning Models of Individual Decision Making to Models of Behavior in Conflict Situations. In *Proceedings of the 2011 International Conference on Behavioral Decision Making.* The Interdisciplinary Center IDC Herzliya, Israel, May 30- June 1, 2011. pp. 4.
20. Gonzalez, C., & Lebiere, C. (2005). Instance-based cognitive models of decision making. In D. Zizzo & A. Courakis (Eds.), *Transfer of knowledge in economic decision-making* (pp. 148–165). New York: Macmillan (Palgrave Macmillan).
21. Gonzalez, C., Lerch, J. F., & Lebiere, C. (2003). Instance-based learning in dynamic decision making. *Cognitive Science, 27*(4), 591–635.
22. Gonzalez, C., & Martin, J. M. (2011). Scaling up Instance-Based Learning Theory to account for social interactions. *Negotiation and Conflict Management Research, 4*(2), 110–128.
23. Hertwig, R., Barron, G., Weber, E. U., & Erev, I. (2006). The role of information sampling in risky choice. In K. Fiedler & P. Juslin (Eds.), *Information sampling and adaptive cognition* (pp. 72–91). New York: Cambridge University Press.
24. Jajodia, S., Liu, P., Swarup, V., & Wang, C. (2011). *Cyber situational awareness: Issues and research.* New York: Springer.
25. Johnson, N. B. (2011). Cyber attacks up 40%, report says. Online article. Federal Times. http://www.federaltimes.com/article/20110403/IT01/104030301/1035/IT01. Accessed March 10, 2012.
26. Juvina, I., Lebiere, C., Martin, J. M., & Gonzalez, C. (2011). Intergroup prisoner's dilemma with intragroup power dynamics. *Games, 2,* 21–51.
27. Lebiere, C. (1999). Blending: An ACT-R mechanism for aggregate retrievals. Paper presented at the Sixth Annual ACT-R Workshop at George Mason University.
28. Lebiere, C., Gonzalez, C., & Martin, M. (2007). Instance-based decision making model of repeated binary choice. In R. L. Lewis, T. A. Polk & J. E. Laird (Eds.), *Proceedings of the 8th International Conference on Cognitive Modeling* (pp. 67–72). Ann Arbor, MI.
29. Lebiere, C., Wallach, D., & West, R. L. (2000). A memory-based account of the prisoner's dilemma and other 2x2 games. In *Proceedings of the International Conference on Cognitive Modeling* (pp. 185–193). NL: Universal Press.
30. Lejarraga, T., Dutt, V., & Gonzalez, C. (2012). Instance-based learning: A general model of repeated binary choice. *Journal of Behavioral Decision Making, 25*(2), 143–153.
31. Lejarraga, T., Dutt, V., & Gonzalez, C. (2010). Instance-based learning in repeated binary choice. Paper presented at the Society for Judgement and Decision Making, St. Louis, MO,
32. Lovett, M. C., Reder, L. M., & Lebiere, C. (1999). Modeling working memory in a unified architecture: An ACT-R perspective. In A. Miyake & P. Shah (Eds.), *Models of working memory: Mechanisms of active maintenance and executive control* (pp. 135–182). New York: Cambridge University Press.
33. Lye, K.-W., & Wing, J. M. (2005). Game strategies in network security. *International Journal of Information Security, 4,* 71–86.
34. March, J. G. (1996). Learning to be risk averse. *Psychological Review, 103*(2), 309–319.
35. Martin, J. M., Gonzalez, C., Juvina, I., & Lebiere, C. (2012). Awareness of interdependence, and its effects on cooperation. Unpublished manuscript under review.
36. Martin, M. K., Gonzalez, C., & Lebiere, C. (2004). Learning to make decisions in dynamic environments: ACT-R plays the beer game. In M. C. Lovett, C. D. Schunn, C. Lebiere & P. Munro (Eds.), *Proceedings of the Sixth International Conference on Cognitive Modeling* (Vol. 420, pp. 178–183). Pittsburgh, PA: Lawrence Erlbaum Associates Publishers.
37. McCumber, J. (2004). *Assessing and managing security risk in IT systems: A structured methodology.* Boca Raton: Auerbach Publication.
38. Rabin, M. (1993). Incorporating fairness into game theory and economics. *The American Economic Review, 83*(5), 1281–1302.

39. Rapoport, A. (1975). Research paradigms for studying dynamic decision behavior. In D. Wendt & C. Vlek (Eds.), *Utility, probability, and human decision making* (Vol. 11, pp. 349–375). Dordrecht, The Netherlands: Reidel.

40. Rapoport, A., & Chammah, A. M. (1965). *Prisoner's dilemma: A study in conflict and cooperation.* Ann Arbor: University of Michigan Press.

41. Rapoport, A., & Mowshowitz, A. (1966). Experimental studies of stochastic models for the Prisoner's dilemma. *System Research and Behavioral Science,* 11(6), 444–458.

42. Roberts, G. (1997). Testing mutualism: A commentary on Clements and Stephens. *Animal Behaviour, 53(6),* 1361–1362.

43. Roth, A. E., & Erev, I. (1995). Learning in extensive-form games: Experimental data and simple dynamic models in the intermediate term. *Games and Economic Behavior,* 8, 164–212.

44. Roy, S., Ellis, C., Shiva, S., Dasgupta, D., Shandilya, V., & Wu, Q. (2010). A survey of game theory as applied to network security. In J. Ralph H. Sprague (Ed.), *Proceedings of the 43rd Hawaii International Conference on System Sciences.* Los Alamitos, CA: IEEE. doi: 10.1109/HICSS.2010.35

45. Salter, C., Saydjari, O. S., Schneier, B., & Wallner, J. (1998). Toward a secure system engineering metholody. In B. Blakeley, D. Kienzle, M. E. Zurko, & S. J. Greenwald (Eds.), *Proceedings of the 1998 Workshop on New Security Paradigms* (pp. 2–10). New York: ACM

46. Shiva, S., Roy, S., & Dasgupta, D. (2010). Game theory for cyber security. In F. T. Sheldon, S. Prowell, R. K. Abercrombie & A. Krings (Eds.), *Proceedings of the Sixth Annual Workshop on Cyber Security and Information Intelligence Research.* New York: ACM. doi: 10.1145/1852666.1852704

47. Schuster, R. (2000). How useful is an individual perspective for explaining the control of social behavior? *Behavioral and Brain Sciences,* 23(2), 263–264.

48. Schuster, R., & Perelberg, A. (2004). Why cooperate? An economic perspective is not enough. *Behavioural Processes,* 66, 261–277.

Chapter 5
Cyber Maneuver Against External Adversaries and Compromised Nodes

Don Torrieri, Sencun Zhu, and Sushil Jajodia

Abstract This article identifies the research issues and challenges from jamming and other attacks by external sources and insiders. Based on the spirit of cyber maneuver, it proposes a general framework to deal with such issues. Central to our framework is the notion of *maneuver keys* as spread-spectrum keys; they supplement the higher-level network cryptographic keys and provide the means to resist and respond to external and insider attacks. The framework also includes components for attack detection, identification of compromised nodes, and group rekeying that excludes compromised nodes. At this stage, much research is still needed on the design of each component as well as integrating these components seamlessly.

5.1 Introduction

Enabled by recent technical trends, including virtualization and workload migration on commodity systems, widespread and redundant network connectivity, instruction-set and address-space randomization, and just-in-time compilers, among other techniques, *cyber maneuver* is envisioned as a viable way of increasing the robustness of imperfect systems and networks under known and unknown

D. Torrieri (✉)
US Army Research Laboratory, 2800 Powder Mill Road Adelphi, MD 20783, USA
e-mail: don.j.torrieri.civ@mail.mil

S. Zhu
Department of Computer Science and Engineering, Pennsylvania State University,
University Park, PA 16802, USA
e-mail: szhu@cse.psu.edu

S. Jajodia
Center for Secure Information Systems, George Mason University, Fairfax,
VA 22030-4422, USA
e-mail: jajodia@gmu.edu

S. Jajodia et al. (eds.), *Moving Target Defense II: Application of Game Theory
and Adversarial Modeling*, Advances in Information Security 100,
DOI 10.1007/978-1-4614-5416-8_5, © Springer Science+Business Media New York 2013

attacks. The motivation behind cyber maneuver is to create, evaluate, and deploy mechanisms and strategies that are diverse, continually shift, and change over time to increase the complexity and costs for attackers, limit the exposure of vulnerabilities and opportunities for attack, and increase system resiliency. Defenses based on cyber maneuver take a strategy afforded to attackers and reverse it to the advantage of defenders. For successful adoption of cyber maneuver techniques, many challenging research problems remain to be solved; e.g., the security of virtualization infrastructures, secure and resilient techniques to move systems within a virtualized environment, automatic diversification techniques, automated ways to dynamically reconstitute and manage the configurations of systems and networks, quantification of security improvement, and potential degradation due to cyber maneuver. See [1] for related articles.

Most proposed cyber-maneuver strategies are expensive and require a large allocation of resources. For example, the Net Maneuver Commander is a prototype cyber command and control system that constantly maneuvers network-based elements preemptively to improve network resiliency [2]. A much simpler strategy, which is particularly useful against jamming, is to use a set of *maneuver keys*. These keys supplement the higher-level network cryptographic keys [3, 4] and, as shown subsequently, provide the means to resist and rapidly respond to jamming and passive attacks. When a network uses spread-spectrum communications, as do most military ad hoc networks, a maneuver key can be a spread-spectrum key that controls the spectral spreading. Cyber maneuver then entails the periodic changing of the maneuver keys.

Frequency-hopping and direct-sequence modulation are the two classic spread-spectrum methods for interference resistance and jamming suppression, both spreading the original signal spectrum into a wider band. Two frequency-hopping transceivers change frequency channels periodically in a synchronized pattern that is unpredictable to attackers. In direct-sequence systems, high-rate sequences called spreading sequences modulate a data sequence so that the receiver knowing the spreading sequence can decode it correctly, whereas others without the knowledge of the spreading sequence cannot. In both cases, the transceivers share a secret key that controls the frequency-hopping pattern or spreading sequence and serves as a maneuver key. The use of spread-spectrum communication systems controlled by long nonlinear sequences generated by the keys is pivotal to their physical-layer security. As long as the keys remain secure, spread-spectrum systems with sufficient processing gains, interleaving, and error-control codes can thwart sophisticated attacks perpetrated by sources located outside the network [5], including repeater jamming against either direct-sequence or frequency-hopping systems [6]. If the *maneuver period*, which is the time between changes of the maneuver keys, is larger than the time it would take an opponent to cryptanalyze an intercepted direct sequence or frequency-hopping pattern and thereby infer the key, then the network is secure against jamming or interception by an external adversary.

However, additional measures are required to cope with the more insidious attacks generated by a network node that becomes compromised. Since it is controlled by a hostile or selfish insider, a compromised node possesses at least some of the maneuver keys. Using these keys, a compromised node may generate false messages or jamming intended for denial of service within a network [7–11]. Until it is identified and isolated from the rest of the network, the compromised node continues to receive the periodic key updates, which enable it to continue its mischief.

A large number of anti-jamming methods have been proposed for both traditional wireless and sensor networks. They can be broadly categorized into jamming-resistant communication with or without shared secrets. Shared-secret methods are more promising, but proposed designs are inefficient. For normal broadcast communication, all nodes initially communicate in one channel, and every message is sent once. When jamming occurs, these schemes work by repetitive transmissions of the same message over multiple channels, and hence higher message transmission redundancy results. The main problem with these designs is that they only rely on the initial secret (either key or channel) assignments to help them survive the jamming, but they do not attempt to recover from compromised secrets.

We advocate a more efficient and secure design that uses periodic rekeying to renew secrets among benign nodes while thwarting external jamming, interception, or eavesdropping. Once the system is restored to normal operation, only one channel (hence one copy of a message) is needed for broadcast communication. Although there is a rich literature on group key management, it is unclear how to integrate the group key management schemes with the jamming-resistant broadcast systems to achieve the above design goal. The main challenges lie in establishing a process for key renewal despite attacks that could originate in multiple, unknown, compromised nodes.

5.2 Related Work

A number of proposed physical-layer solutions to the jamming problem have been developed recently, and here we highlight a few of the most relevant ones. Uncoordinated frequency hopping [12] assumes no pre-existing secret between two parties. There are many frequency channels available for communication. The sender transmits a message multiple times in randomly selected frequency channels, and the duration of transmitting in each channel is T1. The receiver also randomly hops from channel to channel, but the duration of receiving in each channel T2 is much larger than T1. Assuming that not all frequency channels can be jammed simultaneously, a careful choice of parameters enables the receiver to deduce the frequency-hopping pattern.

Uncoordinated direct sequence [12] is a another technique for jamming-resistant broadcast communication without shared keys. Uncoordinated direct sequence is based on two ideas. First, there is a publicly known set of spreading sequences that are used by the system. The sender transmits multiple copies of the message modulated with different sequences randomly chosen from the set. Presumably, jamming may damage some copies but not all. Each receiver records the signals. If a received modulated sequence has not been jammed, the receiver can decode the message correctly and deduce the correct spreading sequence by trial-and-error with all the sequences in the set. While the absence of shared secrets in uncoordinated frequency hopping or direct sequence sounds interesting in theory, such techniques are very inefficient (with repetitive transmissions of the same message using different spreading sequences or frequency-hopping patterns), only provide probabilistic guarantees, and are subverted when senders are compromised nodes.

In [13], a technique is proposed to mitigate jamming against broadcast applications. The technique adopts a binary key tree to denote the relations among the keys that control spreading sequences or frequency-hopping patterns. Every node in the system (mapped to a leaf node in the key tree) knows all the keys in the path from the leaf to the root. When the communication by a sender using a certain key C is jammed, it indicates at least one of the nodes under the source node with that key has been compromised. Then, the sender transmits the message twice, each time using one of the keys that are the children of C to control the spreading sequences or frequency-hopping patterns. If one of the communications is jammed again, the compromised node must be under that subtree, so its child keys will be used for further communications. By this time, a message has been transmitted to three nodes. This process continues until the compromised node is uniquely identified. The main problems with this approach are the excessive message overhead, no attempt to renew the compromised keys, and the impractical requirement for multiple time-consuming sequence or pattern synchronizations

In [14], a transmitter seeks to broadcast a message to multiple receivers such that colluding groups of compromised nodes cannot jam the reception of any other node. The authors propose efficient coding methods that achieve this goal and link this problem to well-known problems in combinatorics. They also link a generalization of this problem to the Key Distribution Pattern problem studied in combinatorial cryptography. Specifically, a transmitter will secretly choose m frequency channels out of a total of M frequency channels and will send the message simultaneously over these m channels. Each node is assigned a subset of these m channels. The goal is that when a limited number of rogue nodes collude, any benign node will at least have one jam-free channel from which to receive messages. However, there are large overhead costs, and success is not guaranteed.

To address this challenge, in [15] a *split and pairing* scheme is proposed for group key establishment under jamming. In the normal case, one group key determines the spreading sequence or frequency-hopping pattern used for group data communication. Since a compromised node possesses the group key, it can also derive the next sequence or pattern and continue jamming. To solve this problem, the remaining group members are divided into two subgroups based on a known function. Because the compromised node cannot jam two subgroups simultaneously and it takes time to switch from one sequence or pattern to another, one of the two groups will receive and share a new group key. Then each node in this subgroup will pair with one node in the jammed subgroup and share the new group key. Finally, without knowing the new group key, the compromised node will be expelled from the group and will not be able to predict the sequence or pattern any longer. The above work is, however, very preliminary. It assumes that the identity of a single compromised node is already known. Clearly, identification of the compromised node is a critical step for the success of group key renewal because otherwise the compromised node learns the new group key. Identifying multiple compromised nodes is more challenging because they can accuse benign nodes and subvert the decision process.

5.3 Proposed Solutions

5.3.1 Communication and Jamming Models

We assume spread spectrum is used for communications among network nodes (as it is in most military ad hoc networks) and the maneuver key is used to control the direct-sequence spreading or the frequency-hopping pattern. If the maneuver key is unknown, an external attacker is generally unable to jam nodes because of the difficulty in transmitting enough power to overcome the processing gain and error-control code of the spread-spectrum system. When an insider physically controls a node, the stored maneuver key becomes known and available for generating spread-spectrum signals. These signals are potentially much more effective as jamming because they will be accepted by any nodes sharing the same maneuver key.

Once synchronization by the victim node to the jamming signal has occurred, that node will be unable to receive legitimate signals. Instead, the node may receive false messages such as misleading routing information or control packets. Other network nodes may be deterred from sending messages because their carrier sensing determines that the channel is busy. If a victim node has previously synchronized with a legitimate spread-spectrum signal, a compromised node will be able to disrupt or corrupt the legitimate reception only if it produces sufficient power to overcome the processing gain. With enough power, the compromised node will *capture* the victim node in the sense that the latter will be forced to resynchronize to the jamming.

5.3.2 The Framework

To effectively use cyber maneuver with maneuver keys against jamming and other attacks by external adversaries and compromised nodes, the following measures are required.

- Maneuver keys must be strong enough to resist cryptanalysis for the duration of the maneuver period.
- Every node needs to detect when it is a victim node subjected to attacks. Detection is a necessary prelude to applying countermeasures against the attacks.
- The nodes or the network operator must have the means to identify the compromised nodes and exclude them from the rekeying process.

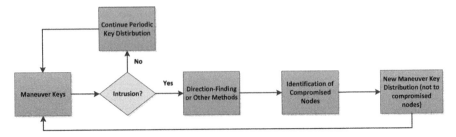

Fig. 5.1 Our framework for attacker identification and maneuver-key updating

- Maneuver keys must be periodically distributed to or independently generated by the legitimate nodes *securely* and *reliably*.

To meet these requirements, we propose the following cyber maneuver framework, as shown in Fig. 5.1. This conceptual block diagram illustrates the process of periodic rekeying, attack detection, identification of compromised nodes, and (triggered) group rekeying. Periodic rekeying is used to thwart cryptanalysis, and it runs even when no attack occurs. When an attack is detected, our system attempts to identify the compromised nodes; finally, a group rekeying is triggered, and new maneuver keys are distributed or generated. Subsequently, we briefly discuss the design of each component as well as its related research challenges that need to be addressed.

5.3.3 Periodic Rekeying

The fundamental method of cyber maneuver for secure wireless communication is the use of a set of maneuver keys in addition to the cryptographic keys (i.e., message encryption keys) used in the network. *A maneuver key is a key that controls network communications and changes periodically with a period shorter than the time it would take for cryptanalysis or subversion of the key.* If the set of maneuver keys is rekeyed or changed periodically with a short enough period, then the main threat will be from a compromised node. Each network node can periodically generate a new maneuver key independently by a cryptographically strong pseudorandom function that precludes prediction of the new key by an adversary.

Since they are to be changed periodically and rapidly, maneuver keys are only required to resist cryptanalysis during the period or interval between changes. Maneuver keys require a key management system for key generation, storage, distribution, and revocation, but the system can be much simpler than the one used for the network cryptographic keys. For example, the maneuver keys can be symmetric rather than comprising public and private keys, and their generation and storage can be done locally. Maneuver keys provide a first line of defense against the compromise of the higher-level network cryptographic keys.

Periodic rekeying alone cannot provide security once a node is compromised because the compromised node can also derive the future keys in the same way as a node that has not been compromised. Hence, a research question is how to do a rekeying that excludes the compromised nodes.

5.3.4   Detection of Attacks

Attacks launched from compromised nodes exploit known network functions, protocols, and shared secrets and are focused against critical network functions such as channel access and routing. An attack can be either an *active jamming attack* or a *passive packet-dropping attack*. A jamming attack may be directed toward the disabling of a node or the sending of false messages to it. Jamming entails the transmission of a signal, and hence is potentially susceptible to direction finding and localization. A passive attack, which is also called *misbehavior*, entails the dropping of packets or failure to forward them when they are routed through a compromised node. Successful passive attacks reduce throughput and increase retransmissions. Although passive attacks are not subject to possible direction finding and localization using received signals, they are less flexible than jamming. Various statistical metrics can be collected by neighboring nodes to potentially identify a compromised node launching passive attacks.

Jamming detection provides knowledge of the presence of jamming in a wireless network. In [11], various measurements are examined to detect jamming including packet-send ratio, packet-delivery ratio, signal strength, and carrier-sensing time. Two other algorithms [16] proposed for detecting jamming are based on bad-packet ratio, packet-delivery ratio, energy-consumption amount, and neighboring-nodes conditions. A sequential technique in [17] detects jamming when an increased number of message collisions are observed during an observation window compared with the previously learned long-term average. Another technique [18] is to select packets dropped per terminal and signal-to-noise ratio as the inputs to a fuzzy inference system that produces a jamming index for each node.

A passive attack or jamming might also be detected above the physical layer of the protocol stack. If a network node finds that it can no longer communicate with a victim node, then it sends a warning message to the victim node by an alternative route (if it exists). When the victim node receives a sufficient number of warnings from other network nodes, then it will know that it is the target of an attack.

5.3.5   Identification of Compromised Nodes

If a successful attack occurs but the compromised node is identified, then the cyber maneuver will truncate the attack's duration by distributing new maneuver keys to benign nodes. These nodes cease to exchange keys with the compromised node, thus preventing it from launching further attacks.

An external adversary or compromised node does not proclaim its own identity, so a network must rely on the observations of a victim node or neighboring nodes to make the inference that an attack has occurred. In attempting to subvert a network, a compromised node has limited options that do not ultimately entail communications between the compromised node and other network nodes. Any communication signal transmitted by the compromised node is susceptible to localization by the network nodes that receive the signal.

Various approaches have been proposed in order to locate the malicious device. Centralized algorithms rely on previously collected measurements, such as node coordinates and the corresponding received signal strength (RSS). RSS-based algorithms can be used to locate a jamming node in principle [19], but the attenuation power law and the shadowing standard deviation [6], which are time-varying in a mobile network, must somehow be measured, and that is generally impractical. Also, these algorithms provide no protection against interference. However, multiple neighbors can use RSS-based algorithms to derive the rough geographical location of a compromised node, thereby restricting the suspect list.

To more accurately determine the location, and hence the identity of the compromised node, a node with multiple antennas can use the DESIC algorithm [20] to find the direction to the source of a jamming signal. The location of the compromised node can be pinpointed by triangulation if the victim node and a neighboring benign node both find directions to the compromised node. Since a compromised node cannot simultaneously transmit normal messages with one direct sequence or frequency-hopping pattern while jamming another, the presence of a node in one group while jamming happens in another group can be used as the proof that the node is benign [21].

Distributed decision-making with multiple compromised nodes can be addressed by adopting the concept of mutual suicide [22]. When a node accuses another one, both of them are revoked temporarily. Later, one is brought back into the network, and it is observed whether an attack resumes. If so, the node is revoked, and the other one is brought back. The method requires a benign node to sacrifice temporarily for the benefit of the group. Clearly, a malicious node can also accuse a benign one, but it does not gain much benefit from doing so.

There are research challenges remaining. First, false positive and false negative localizations could be an issue when localization errors are large. Second, the complexity of identifying multiple compromised nodes is likely to increase at a far higher than linear rate as the number of such nodes increases.

5.3.6 Rekeying After Compromised Node Identification

Once compromised nodes are identified, new maneuver keys should be generated and distributed to all legitimate nodes *securely* and *reliably* when the next maneuver period begins. "Securely" means the compromised nodes do not receive or cannot decode the keys, and "reliably" means all legitimate nodes should receive the new keys even in the presence of jamming. With the same maneuver keys, benign nodes can once again synchronize and communicate. After that, the nodes will continue periodic rekeying.

Although group rekeying has been extensively studied for over a decade, rekeying in the presence of jamming is a new research issue because the channels used for key distribution are themselves susceptible to jamming. The split and pairing scheme was designed for group-key establishment under jamming by a single compromised node, but additional study on its application to jamming by multiple compromised nodes is needed.

5.4 Conclusion and Future Work

This article identified the research issues and challenges from jamming and other attacks by external adversaries and compromised nodes. Based on the spirit of cyber maneuver, it proposes a general framework to deal with such issues. The framework includes components for attack detection, identification of compromised nodes, and group rekeying in the presence of jamming. Some future research is to (1) do detailed designs of each component in the framework, and (2) implement and evaluate both the security and performance of each component as well as the entire system.

Acknowledgements The work of Sencun Zhu was funded by the US Army Research Laboratory under cooperative agreement number W911NF1120086.

References

1. S. Jajodia, A. K. Ghosh, V. Swarup, C. Wang, and X. S. Wang, *Moving Target Defense - Creating Asymmetric Uncertainty for Cyber Threats*, Springer Advances in Information Security Series, vol. 54, 2011.
2. P. Beraud, A. Cruz, S. Hassell, and S. Meadows, "Using cyber maneuver to improve network resiliency," *Proc. IEEE Military Commun.Conf. (MILCOM)*. The work of Sushil Jajodia was performed while he was a Visiting Researcher at the US Army Research Laboratory, 2011.
3. J. V. Merwe, D. Dawoud, and S. McDonald, "A survey on peer-to-peer key management for mobile ad hoc networks," *ACM Computing Surveys*, vol. 39, pp. 1–45, 2007.
4. M. Nogueira, E. D. Silva, A. Santos, and L. C. P. Albini, "Survivable key management on WANETs," *IEEE Wireless Commun.*, pp. 82–88, Dec. 2011.
5. D. Torrieri, *Principles of Spread-Spectrum Communication Systems, 2nd ed.*, Springer, 2011.
6. D. J. Torrieri, "Fundamental limitations on repeater jamming of frequency-hopping communications", *IEEE J. Select. Areas Commun.*, vol. 7, pp. 569–575, May 1989.
7. K. Pelechrinis, M. Iliofotou, and S. V. Krishnamurthy, "Denial of service Attacks in wireless networks: the case of jammers," *IEEE Commun. Surveys Tuts*, vol. 13, 2nd quarter, pp. 245–257, 2011.
8. Y.-S. Shiu, "Physical layer security in wireless networks: a tutorial," *IEEE Wireless Commun.*, pp. 66–74, April 2011.

9. L. Lazos and M. Krunz, "Selective jamming/dropping insider attacks in wireless mesh networks," *IEEE Network*, Jan./Feb. 2011.
10. A. Mpitziopoulos, D. Gavalas, C. Konstantopoulos, and G. Pantziou, "A survey on jamming and countermeasures in WSNs," *IEEE Commun. Surveys Tuts.*, vol. 11, pp. 42–56, 4th quarter, 2009.
11. W. Xu, K. Ma, W. Trappe, and Y. Zhang, "Jamming sensor networks: Attacks and defense strategies," *IEEE Network*, May/June 2006.
12. C. Popper, M. Strasser, S. Capkun, "Anti-jamming broadcast communication using uncoordinated spread spectrum," *IEEE J. Sel. Areas Commun.*, vol. 28, pp. 703–715, June 2010.
13. J. T. Chiang and Y.-C. Hu, "Cross-layer jamming detection and mitigation in wireless broadcast networks," *IEEE/ACM Trans. Networking*, vol. 19, pp. 286–298, Feb. 2011.
14. Y. Desmedt, et al., "Broadcast anti-jamming systems," *Elsevier Computer Networks*, 2001.
15. X. Jiang, W. Hu, S. Zhu, and G. Cao, "Compromise-resilient anti-jamming for wireless sensor networks," *Proc. Twelfth Intern. Conf. Inform. and Commun. Security (ICICS)*, 2010.
16. M. Çakiroğlu, A. T. Özcerit, "Jamming detection mechanisms for wireless sensor networks," *Proc. Third Intern. Conf. Scalable Inform. Systems*, pp. 4:1–4:8, 2008.
17. M. Li, I. Koutsopoulos, R. Poovendran, "Optimal jamming attacks and network defense policies in wireless sensor networks," *Proc. 26th IEEE Intern. Conf. Computer Commun.*, pp. 1307–1315, 2007.
18. S. Misra, R. Singh, S. V. R. Mohan, "Information warfare-worthy jamming attack detection mechanism for wireless sensor networks using a fuzzy inference system," *Sensors*, vol. 10, pp. 3444–3479, 2010.
19. K. Pelechrinis, I. Koutsopoulos, I. Broustis, and S. V. Krishnamurthy, "Lightweight jammer localization in wireless networks: system design and implementation," *Proc. Globecom Conf.*, 2009.
20. D. Torrieri and K. Bakhru, "Direction finding for spread-spectrum systems with adaptive arrays," *Proc. IEEE Military Commun. Conf.*, 2006.
21. H. Nguyen, T. Pongthawornkamol, and K. Nahrstedt, "A novel approach to identify insider-based jamming attacks in multi-channel wireless networks," *Proc. IEEE Military Commun.Conf. (MILCOM)*, 2009.
22. S. Reidt, M. Srivatsa, and S. Balfe, "The fable of the bees: Incentivizing robust revocation decision making in ad hoc networks," *Proc. ACM Conf. Computer and Commun. Security (CCS)*, 2009.

Chapter 6
Applying Self-Shielding Dynamics to the Network Architecture

Justin Yackoski, Harry Bullen, Xiang Yu, and Jason Li

Abstract The static nature of computer networks allows attackers to gather intelligence, perform planning, and then execute attacks at will. Further, once an attacker has gained access to a node within an enclave, there is little to stop a determined attacker from mapping out and spreading to other hosts and services within the enclave. To reduce the impact and spread of an attack before it is detected and removed, semantic changes can be made to several fundamental aspects of the network in order to create cryptographically-strong dynamics. In this chapter, we describe such an architecture designed on top of IPv6 for a wired network enclave. User and operating system impacts are mitigated through the use of a hypervisor, and the dynamics remain compatible with existing network infrastructure. At the same time, an attacker's ability to plan, spread, and communicate within the network is significantly limited by the imposed dynamics.

6.1 Vulnerability of Networks to Attacks

Various types of malicious behaviors are difficult, if not impossible, to prevent using static networks and detection-based techniques. In the most basic case, network scanning and packet sniffing are simple but important ways for an attacker to gather useful information about an enclave. Even an unskilled attacker can scan IP addresses to identify available services and identify potential vulnerabilities. Although the transition to IPv6 creates an address space size that effectively prohibits blindly scanning an entire enclave's address range, by assuming that IPs are assigned sequentially or by observing the IPs used in legitimate network traffic, a "hit list" of in-use addresses can be easily obtained to overcome this [1]. By further

---

J. Yackoski (✉) • H. Bullen • X. Yu • J. Li
Intelligent Automation, Inc., Rockville, MD, USA
e-mail: jyackoski@i-a-i.com; hbullen@i-a-i.com; xyu@i-a-i.com; jli@i-a-i.com

S. Jajodia et al. (eds.), *Moving Target Defense II: Application of Game Theory and Adversarial Modeling*, Advances in Information Security 100, DOI 10.1007/978-1-4614-5416-8_6, © Springer Science+Business Media New York 2013

analyzing traffic, an attacker can focus their efforts on high-value targets within the hit list to target file servers, directory servers, etc. These types of malicious behaviors typically require careful analysis to weed out from legitimate behaviors, making the gradual collection of information about a static network feasible.

Although an enclave typically has a heavily secured perimeter firewall that performs various scanning tasks to protect the network from outside attackers, in practice such a firewall can and will be avoided. With the shift in the motivation of attackers away from notoriety and curiosity towards financial gain and state-sponsored espionage, a specific "zero-day" vulnerability or virus can be reserved specifically for use in a given attack. It is therefore not prudent to rely solely on the hope that a firewall processing millions of packets every minute can perform the analysis needed to meaningfully detect software that exploits a previously unknown attack vector. Even if this were the case, a user with direct access to a node in the enclave can be deceived into directly infecting that node, completely bypassing any network-based detection. According to some estimates, over 25% of modern malicious software is designed to spread via USB [2]. The feasibility and danger of such attacks was demonstrated when a virus from a USB flash drive spread through classified and unclassified military networks [3].

The entire enclave can thus be potentially exposed via a single software vulnerability or an ill-informed act by a user. When an attacker compromises a single node, one should assume that the malicious software has compromised the node's OS and so can monitor all information sent to and from the user and the network. The compromised node can then easily learn the IP addresses of important nodes, record all usernames and passwords observed, hijack connections, etc. Network services requiring use of a typical hardware smart card and PIN system are not immune, as the user's PIN can be easily captured by the compromised OS upon login, allowing the attacker to then authenticate to other nodes and services at will [4].

As a result of these issues, there is disappointingly little stopping an attacker from spreading once inside the enclave. In response to this and other concerns, various security vendors have developed advanced intrusion detection system as well as network switches which scan and selectively block packets transmitted between hosts within the enclave [5]. Such devices rely on two mechanisms for protection. First, users must supply valid credentials which, as discussed, can be easily captured by a compromised OS. Second, any *detected* anomalous behavior results in removal of the PC's access. Obviously malicious behaviors (e.g., worms flooding the network) can often be detected. However, there is always a risk that a determined attacker may use a combination of stolen credentials, knowledge of the detection system's alarm thresholds, and patience to avoid detection-based technology. In fact, if such detection systems are known to be present, this is the logical course for the attacker to take.

6.2   Creating Security Within the Network Architecture

As an alternative to the expensive and imperfect detection of attackers, it is possible to improve network security by instead constantly manipulating the appearance of the network to create a moving target. In this chapter, we focus on introducing dynamics at the network layer. While the dynamics are imposed at the network layer where basic connectivity is provided, the dynamics can be designed to have effects which secure link layer and lower attributes such as the locations and identities of the devices on the network, as well as OS and application layer attributes such as service availability, user authentication, and the network topology.

The goal of these dynamics is to shift the expense of detection to the attacker in several ways. An attacker should be forced to spend significant resources to carefully guide attacks, as packets sent which do not correctly follow the network dynamics allow the attack to be easily and immediately detected. Second, attempts to probe or map the network should reveal a view of the network which is sanitized, ambiguous, and time-varying, making both attack planning and detection avoidance more difficult. Finally, the availability of services should be time-varying based on user needs and credentials, completely but flexibly blocking dangerous network behaviors even if they do follow the network dynamics.

The challenge of achieving such a vision lies in the need to impose dynamics for an attacker while simultaneously hiding the dynamics from existing operating systems, applications, user expectations, routers, switches, and other legacy components. After all, complete security could be achieved by making the network completely unusable for everyone. The difficulty lies in achieving the correct balance. One aspect not to be overlooked when imposing dynamics in this way is that the network's semantic meaning must be preserved at some level of abstraction. As a result, such techniques need to include mechanisms to share information about the dynamics as well as to ensure that those "in the know" regarding the network dynamics are legitimate and authorized.

6.2.1   Examples of Attack Vectors to be Mitigated

To illustrate the types of attack vectors which can be addressed within the network architecture, consider the following representative examples.

6.2.1.1   Network Mapping

In static networks, the presence, connectivity, and locations of individual nodes in the network can be obtained with minimal effort. Ping packets can be sent to some or all possible addresses within the subnet to identify active nodes, followed by a second stage of analysis using traceroutes and other basic probes to create a useful view of the network topology. Further, the IPs of actively communicating nodes can be observed at each endpoint of a connection or at any intermediate point in the network. DNS requests, traffic analysis, host fingerprinting, and other more advanced techniques allow the creation of an effectively complete network map for planning, analysis, and follow-up attacks.

This mapping is possible due to two factors. First, the mapping can be done gradually since networks are rarely re-configured or physically re-organized. Second, each user's and/or node's view of the network is also static and reflects the superset of all required connectivity. In other words, the network infrastructure (i.e. the network administrator) generally grants access in a more coarse-grained manner than is necessary.

6.2.1.2 Worms and Other Automated Malware

Malicious software which automatically spreads and/or operates in the network represents a slightly more sophisticated attack. Such software can be placed into two classes. Software which is not aware that dynamics have been added to the network will inadvertently attempt to access the network in very obviously wrong ways, enabling fast and accurate detection. Inevitably, malware authors will create a dynamic-aware class of malware that does not fall so easily into detection traps and should instead be prevented from achieving its goal before any alarm is raised. This class of software is effectively an Advanced Persistent Threat.

As with network mapping, malware can operate effectively because of the overly restrictive static access policies often in place. For malware with distributed command and control architectures, the network dynamics should impose challenges to the formation and maintenance of peer-to-peer membership and other information sharing.

6.2.1.3 Advanced Persistent Threats

An emerging and significant threat category is the Advanced Persistent Threat (APT), which includes advanced semi-autonomous malicious software or a human with a malicious intent manually accessing the network over a relatively long period of time. In both cases, a node in the network enclave is typically infected and used remotely as an entrypoint for further spread. An APT has several goals including to spread to higher-value nodes, to collect and exfiltrate information, and to command and control the infected nodes.

An APT acts in ways described in the above discussions, but further needs to achieve specific malicious purposes by carefully identifying nodes to spread to which have the highest risk-reward tradeoff, and to establish purpose-specific communication pathways within and out of the network.

6.2.1.4 Security Through Obscurity

As a final note on attack vectors, it is important to consider that in many applications it is not sufficient for the network to be protected *unless* the attacker correctly makes a one in a trillion guess. On large timescales and with sufficient computing power, such guesses may be possible. Any dynamics should therefore either be secure from a cryptographic standpoint or their limitations well understood.

6.2.2 Desired Properties of a Network Architecture

To address these problems, the dynamics added to a network architecture should possess the following attributes.

- *Transparency* from the perspective of the user, application, and OS.
- *Cryptographically strong* prevention of undesired network actions.
- *Dynamic availability* of services as appropriate.
- *Concealment* of the addresses and identities of nodes and services.
- *Fault-tolerant, distributed operation* which does not cause the network to be fragile or possess added points of failure.
- A *Combination* of multiple techniques.
- *Compatibility* with existing network security and management technology.

6.2.2.1 Possible Advantages Over Existing "Static" Security Techniques

It is important to note that some of these properties could instead be achieved using a well-configured VPN, firewall, and other existing network security paradigms. However, using the approach described in this chapter provides significant additional capabilities which a VPN and/or firewall cannot offer. In general, firewalls and VPNs are not intended to provide protection to communication occurring solely within an enclave, and detection of attacks occurring within an enclave is non-trivial. The most significant additional capabilities which can be achieved are as follows.

- *Protection against information gathering*—A variety of techniques can be used to prevent the gathering of useful information about the network. A VPN does hide the true source and destination while the packet is traveling over the ciphertext segment of the network. However, a VPN does not prevent the endpoints from knowing each other's identity, protect communication within the enclave, or effectively conceal the structure of the network. Similarly, a firewall does not prevent a user from discovering information about accessible services.
- *Protection against compromised PCs*—A firewall or VPN cannot determine whether the user's credentials are being misused.
- *Fine-grained security controls*—While a firewall is highly configurable, it is difficult to correctly act on a given packet. If the sender of each individual packet

can be cryptographically verified due to the fact that they follow the dynamics appropriately, this simplifies configuration and reduces the chance of unintentionally granting access, which is often a source of security vulnerabilities.

- *Elevation mechanism and audit trail*—A restrictive policy can be provided by default if the dynamics mechanisms provide a verifiable mechanism to achieve further network access when needed. On a packet by packet basis, the sender can then be verified and a determination made as to whether the packet's sender has already provided the necessary authentication. Further, if a packet or connection is found to be malicious in some way, the offender's identity can be proven.
- *Security by default*—The network does not allow packets to be sent through it unless they follow the dynamics correctly. Since these dynamics can only be calculated by a valid and properly configured agent, any problem with the low level setup of the network will result in an unintentional *limiting* of access to the network. In contrast, a misconfigured firewall or VPN can easily cause an unintentional increase of access.

6.3 A Case Study

To illustrate some of the security benefits achievable using these types of network architecture level dynamics, we describe the Self-shielding Dynamic Network Architecture (SDNA) [6]. By inserting a hypervisor within each network node, SDNA makes the network's appearance dynamic to observers while at the same time retaining the necessary semantics for transparency to legitimate users. This is illustrated in Fig. 6.1. When in the network "below" SDNA, packets appear to be traversing a dynamic network. On the OS, application, and user side of SDNA, the dynamics are concealed only to the extent necessary to provide an abstract but semantically valid view of the network.

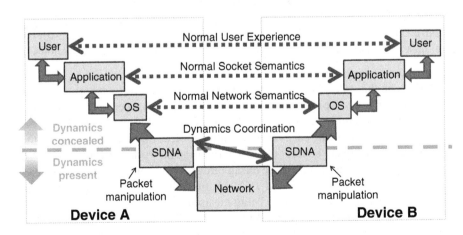

Fig. 6.1 Semantic view of SDNA's dynamics

The objective of SDNA is to manage the competing goals of simultaneously securing the OS from the network while providing the user access to needed services. This architecture provides a way to disrupt the planning, effectiveness, and spread of attackers within a network. Through cryptographically-secure mechanisms, dynamics are added to the network that prevent the attacker from gathering and acting on information about the network. SDNA is comprised of a combination of existing networking techniques, hypervisor technology, Common Access Card (CAC)-based authentication, and IPv6. This approach simplifies the deployment and management of SDNA within existing networks and reduces the technical risk involved. Incremental deployment supporting legacy (either IPv4-only or SDNA-unaware IPv6) systems can also be supported using standard IPv6 transition mechanisms.

6.3.1 SDNA Overview

The SDNA architecture allows multiple security mechanisms to be coherently combined, while also ensuring that the OS cannot access the network (or vice versa) without proper credentials. Introducing multiple types of dynamics into the network has the potential to create conflicting or destructive effects which might result in a net decrease in security. By integrating these dynamics into a common architecture, the protection of multiple mechanisms can instead be constructively combined to enhance overall security. Network administrators and existing security technologies (IDSs, audit trails, etc.) can also continue to understand the network (see Sect. 6.4.2), with the added benefit of cryptographically verified information provided by SDNA.

The hypervisor transparently rewrites packets entering and exiting the OS to prevent the OS from observing real addresses in the network. To prevent network sniffing outside the OS, packets travel through one or more intermediate nodes before reaching their final destination. Each node's hypervisor uses existing certificates and authentication mechanisms to independently verify the source of each packet and connection. The network's behavior and appearance can then be varied based on the result of this verification. The availability of each service can be changed in response to the proper credentials with a provable audit trail, giving network administrators the ability to enact highly restrictive but flexible security policies. Further, misconfiguration and accidental exposure of a node or service to the rest of the network is much more difficult because two nodes must correctly follow the network dynamics in order to successfully communicate. As a result of SDNA, the OS is unable to observe the credentials supplied by the user or connect to arbitrary destinations without the user's knowledge. Equally important, the OS is not able to access the network or to connect to a service without the user first providing a cryptographically secure credential directly to the server via SDNA.

In SDNA, a hypervisor is placed on each node in the network to facilitate SDNA's action as an intermediary between the network and the OS or *Guest*, as shown in

Table 6.1 Division of IPv6 address space to create SDNA bits

Name	Network ID	Host ID	SDNA bits
Length	48 bits	40 bits	40 bits
Example	fef0:0000:0000:	0000:0000:22	00:0000:0001

Fig. 6.1. The Guest resides in one virtual machine (VM), while another contains the SDNA functionality, which we refer to as the *SDNA Entity*. To achieve scalability and avoid single points of failure, each SDNA Entity operates independently to process packets and directly coordinates with other relevant SDNA Entities to facilitate communication between Guests. In terms of management, the normal operations of SDNA are automatic and no centralized control or input from the network administrator is required. The SDNA Entity is the core component of our design and contains all of the dynamics mechanisms comprising SDNA. In a typical case, packets generated by the Guest are received by the SDNA Entity which uses the packet's metadata and a cryptographic key to rewrite the IP addresses before transmitting them to the network. Upon receiving packets from the network, the SDNA Entity verifies the packets and then rewrites the IP addresses in the packets before providing the packets to the Guest. In this way, the SDNA Entity is able to introduce arbitrary dynamics into the network while at the same time modifying packets traveling to and from the Guest so that the dynamics are transparent. We assume that a shared key exists between the SDNA Entities of two communicating Guests. The creation of the necessary keys occurs during an exchange process initiated by the user and/or OS. This key allows sender verification of each packet, providing the ability to perform host- and service-level availability depending on the result of the verification. Packets failing the verification can be redirected to a honeypot or dropped.

6.3.1.1 Architecture and SDNA Bits

Before discussing specific dynamics mechanisms included in SDNA, it is important to consider how dynamics are introduced into the network. Specifically, changing the *appearance* of the network must involve changing each node's address since this is the fundamental attribute of each node. At the same time, existing OSs and network hardware (routers, switches, etc.) must still function in the presence of any dynamics to ensure that SDNA can realistically be deployed. Within an enclave, two addresses—the link-layer (MAC) address and the network-layer (IPv4 or IPv6) address—are used for communication and routing/switching. We consider dynamics based on the IPv6 addresses due the fact that an IPv4-only solution is not future-proof and any IPv4 traffic can be easily handled via an IPv6 transition mechanism.

We divide the 128 bit IPv6 address as shown in Table 6.1. Giving a portion of the IPv6 address space a special semantic meaning has been suggested in [7]. According to current IPv6 allocation policies [8], each *site* is assigned a /48 subnet, making the Network ID the first 48 bits of the IPv6 address. A site is analogous to

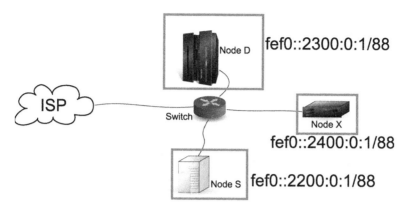

Fig. 6.2 Example network topology

an enclave and a /48 allocation is recommended for allocation to each business by their Internet Service Provider (ISP), with larger allocations presumably possible for very large institutions and smaller allocations for home users. We use the final 40 bits of each IPv6 address for SDNA and refer to them as the *SDNA bits*. This leaves 40 bits in the middle for the Host ID. In other words, each site can have 2^{40}, or almost 1.1 trillion, hosts. Thus, SDNA's use of the final 40 bits of IPv6 has no foreseeable negative impact on the amount of addresses available for a practical enclave network. By combining the last 40 bits of both the source and destination IP address into a single logical field, each packet has 80 SDNA bits. The SDNA bits are calculated by securely hashing the packet's metadata, the current time, and a shared key. In our implementation, we use a SHA1 hash of the IP header (excluding the SDNA bits and TTL) as well as the majority of the ICMP/TCP/UDP headers (excluding the checksum).

For example, a host might be assigned fef0::2200:0:0/88 (i.e. fef0::2200:0:0 through fef0::22ff:ffff:ffff). When a packet enters the SDNA entity from the Guest, SDNA sets the correct values for the SDNA bits. When a packet enters the SDNA entity from the network, the SDNA bits are cryptographically verified. If the SDNA bits are correct, SDNA sets the bits to match the Guest's IP address and normal communication takes place. If they are invalid, SDNA modifies the IP addresses to reroute them to a honeypot or other monitoring device.

6.3.1.2   Service Redirection

Within an enclave, nodes normally directly communicate without going through a router, allowing an attacker to easily observe whom a given node is communicating with. For example, in the topology in Fig. 6.2 an attacker observing all of Node S's packets could determine that S is communicating with D. We thus require that all communications go through one or more *intermediate nodes*. Intermediate nodes (e.g., Node X) are not simply routers, but also rewrite traffic between various endpoints at their request to conceal the endpoints' identifies both from each other and from potential attackers. For maximum security, the intermediate nodes should be dedicated PCs with no Guest and placed on a separate LAN segment to prevent compromise by an attacker. Further, all traffic must pass through an intermediate node, direct communication between communicating nodes is not permitted by the network configuration.

This approach requires coordination between nodes in order to establish the appropriate configuration with respect to communication via the intermediate nodes. For communication within an enclave, the coordination needed for Service Redirection is completely distributed and involves only the source, destination, and any chosen intermediate nodes. To allow external communication into an enclave (either from a third party or from another enclave with SDNA-protected hosts), the node hosting a given service must pre-select one or more initial intermediate nodes and then inform the enclave's DNS server of the intermediate nodes' IP addresses. The external node then simply looks up the address using DNS, unaware that redirection is occurring. Thus no coordination is needed between enclaves.

6.3.1.3 Key Exchange and Authentication

A shared secret key exists between the SDNA Entities of each pair of communicating Guests to allow them to create and validate the appropriate dynamics. The presence or lack of this key, along with any associated security policies, allows the SDNA Entities to dynamically control the availability of the network and to drop or transparently redirect packets to a honeypot as appropriate.

In practice, it is easy for two SDNA Entities to authenticate each other in an automatic and cryptographically secure way; the difficulty lies in determining that the packet was generated by a legitimate user and not an attacker which has compromised the Guest. To this end, we suggest the use of standard best practices such as the presence of a hardware card combined with a PIN or one-time use token. Policy is an important aspect, in that high risk behaviors should require more stringent verification.

It is critical that the request for user input as well as the user's response be sent and received directly to/from the SDNA Entity as shown in Fig. 6.3 and that it is not observable to the Guest. If the Guest is compromised, the user's input could be replayed or the user could be tricked into providing authentication data for one service while the guest actually connects to another service using the provided authentication data.

6.3.1.4 Address and Naming Dynamics

SDNA conceals the identity of other nodes within the enclave from each Guest. In addition to the DNS reply modification discussed above, each SDNA Entity monitors DNS responses coming to its own Guest and replaces any IPs with

Fig. 6.3 For security, the
user authentication must
bypass the OS

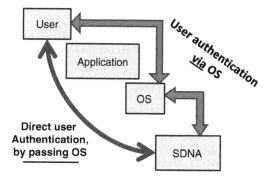

Token IPs. The Token IP is generated by the SDNA Entity and acts as an identifier which allows the SDNA Entity to later look up the correct IP(s) associated with that Token IP in the SDNA Entity's internal database in order to convert back when the Guest sends a packet to that Token IP. The Guest uses the Token IP for storing connection state, thus the Token IP must remain valid for a Guest's entire connection. When the Guest initiates a connection to a Token IP, the SDNA Entity replaces the Token IP with the real IP. Similarly, the SDNA Entity at the destination substitutes another Token IP for the source's IP, so that neither the Guest initiating the connection nor the Guest listening for connections knows each other's real IP.

6.3.1.5 Honeypot

In cases where the verification of a packet's SDNA bits fail, the packet is transparently redirected to the honeypot via an intermediate node. The honeypot's SDNA Entity provides the honeypot with Token IPs, in the same way as the Guest, so that the honeypot can operate without understanding any network dynamics. By automatically redirecting packets with invalid SDNA bits to the honeypot, the attacker must expend significantly more effort trying to access desired services. The honeypot prevents an attacker from effectively conducting hash attacks on the SDNA bits for a packet since the response is similar in both cases. The honeypot allows SDNA to create fake services that make it unclear to attackers whether they have been authorized to use a given service. The attacker must then carefully analyze the responses from the service to determine whether the attack has actually been successful.

6.3.2 Example of Dynamic Network Operation

In this section, we show examples of SDNA's operation in a testbed network. We have designed and implemented SDNA using a Linux-based SDNA Entity running on laptops and network devices, including key techniques to introduce dynamics

```
yackoski@guest-s:~$ ping6 guest-d
PING guest-d(fef2::1) 56 data bytes
From sdna icmp_seq=1 Destination unreachable: Administratively prohibited
From sdna icmp_seq=2 Destination unreachable: Administratively prohibited
^C
--- guest-d ping statistics ---
2 packets transmitted, 0 received, +2 errors, 100% packet loss, time 1003ms

yackoski@guest-s:~$ telnet guest-d
Trying fef2::1...
telnet: Unable to connect to remote host: Permission denied
yackoski@guest-s:~$ wget http://guest-d/
--2010-08-23 11:01:58--  http://guest-d/
Resolving guest-d... fef2::1
Connecting to guest-d|fef2::1|:80... failed: Permission denied.
```

Fig. 6.4 Guest S attempting to access the network without authentication

into the network, with the goal of preventing an attacker from learning about and spreading through the network. We show SDNA's basic operation in a testbed network using the topology and assigned IP addresses shown in Fig. 6.2 with both an SDNA Entity and Guest present on nodes S and D, and only an SDNA Entity present on intermediate node X.

6.3.3 Normal Operation

In the typical case, a User S on Guest S wishes to access some service on Guest D. When the user has not been authenticated to the SDNA Entity, we have configured the SDNA Entity to politely inform the Guest and user that authentication is required. This is shown in Fig. 6.4, since the user is initially not authenticated to the desired service. By using the ICMPv6 protocol already supported by many operating systems and applications, the SDNA Entity can immediately notify the specific application making a connection request. Note that the SDNA Entity on Node S has modified the DNS response received by Guest S to provide a Token IP address of fef2::1 instead of Guest D's real IP address.

The user on Guest S must then authenticate to the SDNA Entity. This authentication can be one-time, or can be specified on a per-service or per-connection basis. The packet capture in Fig. 6.5 shows that the key exchange begins with an SSL-protected bootstrap phase between node S and X to setup the key (packets 69–88), followed by data communication to node D via node X using the hashed key to set the SDNA bits (packets 91 onward).

As shown in Fig. 6.6, after authentication the user on Guest S can connect to Guest D's HTTP service (note the 404 error indicates a successful *connection*, although no webpage exists). When connecting to Guest D's ssh service using the guest-d domain name (perhaps for malicious purposes), the user is instead redirected

No.	Source	Destination	Protocol	Info
69	fef0::2200:0:2	fef0::2400:0:0:2	TCP	33454 > 4433 [SYN]
70	fef0::2400:0:0:2	fef0::2200:0:2	TCP	4433 > 33454 [SYN,
71	fef0::2200:0:2	fef0::2400:0:0:2	TCP	33454 > 4433 [ACK]
72	fef0::2200:0:2	fef0::2400:0:0:2	TCP	33454 > 4433 [PSH,
73	fef0::2200:0:2	fef0::2400:0:0:2	TCP	33454 > 4433 [FIN,
74	fef0::2400:0:0:2	fef0::2200:0:2	TCP	4433 > 33454 [ACK]
77	fef0::2400:0:0:2	fef0::2200:0:2	TCP	4433 > 33454 [PSH,
78	fef0::2200:0:2	fef0::2400:0:0:2	TCP	33454 > 4433 [ACK]
81	fef0::2400:0:0:2	fef0::2200:0:2	TCP	4433 > 33454 [PSH,
82	fef0::2200:0:2	fef0::2400:0:0:2	TCP	33454 > 4433 [ACK]
87	fef0::2400:0:0:2	fef0::2200:0:2	TCP	4433 > 33454 [FIN,
88	fef0::2200:0:2	fef0::2400:0:0:2	TCP	33454 > 4433 [ACK]
91	fef0::22a8:1647:e4ee	fef0::2400:2855:f4a1:	TCP	36800 > 5533 [SYN]
94	fef0::2400:2800:0:2	fef0::2300:0:2	TCP	36800 > 4433 [SYN]
95	fe80::215:6dff:fec5:c639	ff02::1:ff00:2	ICMPv6	Neighbor solicitat
98	fef0::2400:28eb:d9f0:c6d	fef0::22e0:3c86:c7cf	TCP	5533 > 36800 [SYN,
101	fef0::2297:2e77:b830	fef0::2400:2833:e266:	TCP	36800 > 5533 [ACK]
104	fef0::2215:5b72:d344	fef0::2400:2859:e16d:	TCP	36800 > 5533 [PSH,
105	fef0::2238:6a40:db0e	fef0::2400:2855:e9fe:	TCP	36800 > 5533 [FIN,
106	fef0::2400:2897:dd79:858	fef0::2299:92b0:6bc8	TCP	5533 > 36800 [ACK]
113	fef0::2400:28ca:b39:eac8	fef0::2266:f268:ba2d	TCP	5533 > 36800 [PSH,
116	fef0::2236:a4cb:a96b	fef0::2400:28f1:ed88:	TCP	36800 > 5533 [ACK]
119	fef0::2400:2862:7120:90b	fef0::22ff:8358:c5da	TCP	5533 > 36800 [FIN,
120	fef0::228c:8c1:b042	fef0::2400:2874:8543:	TCP	36800 > 5533 [ACK]

Fig. 6.5   Packet capture of key exchange between S and D

```
yackoski@guest-s:~$ wget http://guest-d/
--2010-08-23 11:10:54--  http://guest-d/
Resolving guest-d... fef2::1
Connecting to guest-d|fef2::1|:80... connected.
HTTP request sent, awaiting response... 404 Not Found
2010-08-23 11:10:54 ERROR 404: Not Found.

yackoski@guest-s:~$ ssh guest-d
yackoski@quest-d's password:
Last login: Mon Aug 23 15:08:32 2010 from fef1::1
yackoski@cross2:~$ █
```

Fig. 6.6   Guest S connects to Guest D via HTTP but reaches the honeypot when connecting to the ssh service

Fig. 6.7   Guest S connects to Guest D's ssh service after separate authentication

```
yackoski@guest-s:~$ ssh ssh.guest-d
yackoski@ssh.guest-d's password:
Last login: Mon Aug 23 07:43:57 2010 from fef2::3
yackoski@guest-d:~$
```

to the honeypot as indicated by the hostname cross2. To access the ssh service, which is given a more restrictive policy, the user must authenticate to the separate domain ssh.guest-d which the SDNA Entity translates into a separate Token IP.

After successful authentication, the user is then able to connect to Guest D's ssh service as shown in Fig. 6.7.

```
IAI-OpenWrt-26 ~ :) # ping fef0::2300:0:1
PING fef0::2300:0:1 (fef0::2300:0:1): 56 data bytes
64 bytes from fef0::2300:0:1: seq=0 ttl=64 time=0.715 ms
^C
--- fef0::2300:0:1 ping statistics ---
1 packets transmitted, 1 packets received, 0% packet loss
round-trip min/avg/max = 0.715/0.715/0.715 ms
IAI-OpenWrt-26 ~ :) # ifconfig eth1 add fef0::2400:2994:6e26:3e07/48
IAI-OpenWrt-26 ~ :) # ip -6 r get fef0::23a7:6508:6fbd
fef0::23a7:6508:6fbd from :: via fef0::2300:0:2 dev eth1  src fef0::2400:2994:6e26:3e07
it 0
IAI-OpenWrt-26 ~ :) # ping fef0::23a7:6508:6fbd
PING fef0::23a7:6508:6fbd (fef0::23a7:6508:6fbd): 56 data bytes
64 bytes from fef0::23a7:6508:6fbd: seq=0 ttl=64 time=0.749 ms
^C
--- fef0::23a7:6508:6fbd ping statistics ---
1 packets transmitted, 1 packets received, 0% packet loss
round-trip min/avg/max = 0.749/0.749/0.749 ms
IAI-OpenWrt-26 ~ :) # ssh yackoski@fef0::23a7:6508:6fbd
yackoski@fef0::23a7:6508:6fbd's password:
Last login: Mon Aug 23 15:18:04 2010 from fef1::2
yackoski@cross2:~$
```

Fig. 6.8 An attacker using observed addresses and SDNA bits is redirected

```
fef0::2400:0:0:2.0-fef0::2300:0:1.0 Packet received
 looking up key for fef0::2300:0:0 - fef0::2400:0:0:0
fef0::2400:0:0:2.0-fef0::2300:0:1.0 no key found, going to send to honeypot
fef0::2400:0:0:2.0-fef0::2300:0:1.0 packet from network, verifying
fef0::2400:0:0:2.0-fef0::2300:0:1.0 No valid key or session found
fef0::2400:0:0:2.0-fef0::2300:0:1.0 REDIRECTING TO HONEYPOT
fef1::2.0-fef0::2300:0:2.0 rewriting bits in src/dest
---------------------------------------------------------
fef0::2400:2994:6e26:3e07.51736-fef0::23a7:6508:6fbd.22 Packet received
 looking up key for fef0::2300:0:0 - fef0::2400:2900:0:0
fef0::2400:2994:6e26:3e07.51736-fef0::23a7:6508:6fbd.22 found key
fef0::2400:2994:6e26:3e07.51736-fef0::23a7:6508:6fbd.22 packet from network,
fef0::2400:2994:6e26:3e07.51736-fef0::23a7:6508:6fbd.22 REDIRECTING TO HONEYPOT
fef1::4.51736-fef0::2300:0:2.22 rewriting bits in src/dest
```

Fig. 6.9 The SDNA Entity easily detects the attacker's misbehavior

6.3.3.1   Impact on An Attacker

Suppose an attacker somehow determines Node D's real IP is fef0::2300:0:1 and attempts to communicate using this address. As shown in Fig. 6.8, the attacker's pings are successful, however as seen in Fig. 6.9 the first packet does not match any authenticated session (based on the attacker's source IP) and is redirected to the honeypot. Suppose the attacker also sniffs a packet with the SDNA bits set and configures the network stack to send packets to node D using fef0::23a7:6508:6fbd as the source IP and fef0::2400:2994:6e26:3e07 as the destination to re-use these observed SDNA bits. Again, we can see in the lower portion of Fig. 6.9 that a packet is received by D with these addresses however the packet has different metadata (both timestamp and TCP sequence number) resulting in a mismatch of the SDNA bits. This conclusively detects the misbehavior, resulting in redirection to the honeypot.

6.4   Analysis

6.4.1   Security Analysis

We briefly consider the major security risks both to the enclave and to SDNA. There are three basic groups of actions that an attacker may attempt: (1) gathering information about the enclave, (2) accessing services within the enclave, and (3) attacking SDNA itself.

SDNA uses a combination of hypervisor technology, Token IPs, Service Redirection, and SDNA bits to prevent an attacker from observing useful information. Each Guest cannot observe the real IP address of other nodes or services since the SDNA Entity replaces the real IP with a Token IP for all packets coming into the Guest. An attacker outside of a Guest, which has either inserted new hardware or compromised a node not protected by an SDNA Entity, can however observe the IPs in packets sent through the enclave. Due to service redirection, an attacker can only learn the real IP of an endpoint by observing packets sent between the last intermediate node and the endpoint. This requires the attacker to compromise either an intermediate node or the network infrastructure used for communication between the endpoint and intermediate nodes. Following current security best practices of combining servers on a separate network segment and also placing dedicated intermediate nodes (which have only an SDNA Entity and no Guest) on this segment, such an attack is unlikely. This prevents an attacker from directly targeting the SDNA Entity of a high-value service. Finally, assuming an attacker does observe the real IP of a node (including the SDNA bits valid for a sequence of packets) this information cannot be used by the attacker. Replay attacks are prevented by including the current time in the hash used to calculate the SDNA bits, and any changes in a packet's metadata (ports, IPs, sequence number, etc.) change the correct value for the SDNA bits. Thus, knowing the SDNA bits for an observed packet gives the attacker no useful information. The attacker must be able to calculate the correct SDNA bits in order to initiate a useful connection, which requires the shared key known only to the relevant SDNA Entities.

An attacker cannot access services protected by an SDNA Entity without calculating the correct SDNA bits. Assuming SDNA Entities themselves cannot be compromised via a hypervisor exploit, the only method available to an attacker is to compromise a Guest and then trick the SDNA Entity for that Guest into thinking the attacker is a legitimate user. Note that an attacker's software may be running on a Guest at the same time a legitimate user is logged in to the Guest. The attacker can then access any services which the legitimate user has access to since it is difficult for the SDNA Entity to determine the intent or origin of packets received from the Guest. An attacker on a compromised Guest has two hurdles created by SDNA which must be overcome. First, for infrequently accessed or highly sensitive services, the user is required to provide input to the SDNA Entity on a per-connection basis in the form of a PIN, challenge/response, or one-time use token. Second, the user can also be required to enter a secret domain name or other value into the SDNA Entity directly. The SDNA Entity can then generate a Token IP that, while usable to the Guest, cannot be associated to a specific service. Since these values are input directly into the SDNA Entity (via hardware, i.e. the keyboard is directly connected to the hypervisor), there is no way for a compromised Guest to enter such information into the SDNA Entity even if the user mistakenly reveals it to the Guest.

Finally, SDNA itself can be the target of several types of attacks. An attacker may attempt to calculate valid SDNA bits by directly attacking the hash function. With 80 SDNA bits, an attacker must perform roughly 2^{40} operations in order to guess a valid hash. To access a service, the SDNA bits for a series of packets must be calculated. TCP connections require at least 3 packets to initiate. Protocols using TLS, IPSec [9], and other secured channels typically require 10–15 packets for connection setup. This means that an attacker must perform at least 2^{120}–2^{400} operations, which is beyond the capabilities of even a large scale botnet [10]. Due to SDNA's honeypot, the attacker also cannot easily determine whether a guess is correct, further increasing the effort needed. Instead of modifying the SDNA bits, the attacker can modify the payload of an ongoing connection. To counter this, the hash calculation used can include the entire payload. However, SDNA cannot prevent the payload from being modified within the Guest.

6.4.2 Usability Analysis

In terms of user impact, SDNA does not require a significant change in user behavior. The user inputs the information directly to the SDNA Entity instead of entering credentials into the Guest, a change which is essentially transparent to the user. An improperly authenticated user may be directed to the honeypot and become confused. While this is the intent of the honeypot, the user's SDNA Entity can warn the user of such situations depending on policy. The significant change potentially visible to the user is that all IP addresses observed are Token IPs and not the real IPs. Due to the significant length of IPv6 addresses, it follows that users will rely more heavily on domain names and increasingly avoid entering IP addresses directly. Since SDNA transparently preserves the semantics of domain names, a normal user will thus be unaffected. However some applications, in particular ssh, may still use IP addresses as a unique identifier and may become confused if the Token IP for a particular service changes. One way to address this is to always provide the same Token IP for such services, for example by publishing a second IP via DNS that the SDNA Entity always uses as the Token IP.

For network administrators, common diagnostic tools such as ping continue to work. To assist in the troubleshooting of advanced networking problems, we suggest adding self-diagnostic functionality to the SDNA Entity (e.g., the ability to instruct the SDNA Entity itself to perform a ping/traceroute) which can then be activated by the network administrator when a problem occurs.

Existing network security components such as IDSs, logging, etc. are also affected by the dynamics introduced by SDNA. Since SDNA only varies a node's IP address within a defined range, the SDNA bits in each address can be simply set to zeros before the packets are analyzed by such components. Existing tools however do not understand SDNA's use of service redirection. Intermediate nodes in SDNA act as a natural collection point for traffic, thus we suggest having intermediate nodes provide a modified copy of all packets to such security components. For example, an intermediate node X forwarding traffic between nodes S and D receives packets sent from S to X and modifies the packet before forwarding to make it appear to be from X to D. Having all the necessary knowledge, the intermediate node should then provide the security components with a single packet whose header is constructed as if the packet was actually sent from S to D. Using this technique combined with clearing the SDNA bits, existing security components can be accommodated without modifications.

6.4.3 Feasibility Analysis

We have specifically designed SDNA to require minimal effort to add to an existing network. We acknowledge that arguably higher security could likely be achieved through different design decisions; however the practical benefit of such additional security does not outweigh the decreased feasibility and user-friendliness of the other alternatives we have explored. There are no significant technical barriers preventing our current SDNA implementation from being adopted in a real network. SDNA's dynamics, including variations in the IP addresses of nodes, can be added to a network with little or no performance penalty and without requiring an upgrade of network switches and other infrastructure. SDNA relies upon existing user authentication paradigms and requires no modification to the Guest's OS. The primary tasks required to add SDNA to a network involve adding the SDNA Entity and hypervisor to each host, enabling IPv6 in the network if not already enabled, and modifying the network configuration. For IPv4 hosts and networks, standard IPv6 transition mechanisms can be either used separately or incorporated into the SDNA Entity to completely hide IPv6 from the Guest.

On initial setup, each host in an SDNA-protected enclave is then assigned a /88 subnet and this information is provided to the Layer 3 switch. In other words, the network administrator simply instructs any network hardware to treat each node as if it were an entire /88 subnet instead of a single node. In the case of a regular (Layer 2) switch, no switch configuration is required. Normal routing and MAC address resolution protocols can then be used without modification, and the ARP (Address Resolution Protocol) tables only require one entry per node. Two addresses within each /88 range are reserved. The 00:0:1 suffix (e.g., fef0::2200:0:1) is assigned to the Guest and is the default IP used to reach a given node, while the 00:0:2 suffix (e.g., fef0::2200:0:2) is assigned to the SDNA Entity. To overcome the "chicken and egg" problem of establishing keys in order to set and verify the SDNA bits, the SDNA Entities allow packets to be received on these two addresses only for the purposes of exchanging keys.

It is important to note that SDNA modifies IP addresses, and not MAC addresses. Rapid modification of MAC addresses would introduce potentially significant performance and security issues with current switches whose limited Content Addressable Memory (CAM) tables would quickly overflow. Further, the use of intermediate nodes for all communication hides the other nodes' MAC addresses from each other. The network hardware (i.e. switches) must still support any variation in the IP addresses.

We consider three basic types of switches. First, consider a "low end" switch for consumers or small businesses acting based on MAC addresses. Since we do not vary the MAC address, such a switch is not affected by SDNA. Second, consider a Layer 3 switch, more commonly used in enclaves of a hundred or more nodes. Such a switch must be configured with a range of IP addresses for each port. For this reason we suggest only varying the IP addresses within a fixed range quasi-statically configured on the switch, avoiding incompatibility issues. Normal Layer 3 switches, depending on their configuration, maintain a cache (similar to CAM) mapping a previously seen IP address to a given port. If no entry in the cache is found, the switch's configuration is consulted, which takes additional processing. Since SDNA varies the IP addresses, in scenarios with certain specific switch configuration, design, and CAM structure, nearly all lookups via this cache will miss, requiring the configuration to be consulted. For switches carrying low bandwidth, however, the performance impact is not significant. Finally, consider a "high end" Layer 3 switch in a large enclave. These switches are designed for high throughput and act directly based on the prefixes configured by the administrator with no caching. Using techniques such as Cisco Express Forwarding (CEF), IP prefix matching can be done at line speed [11] by enterprise-level switches. SDNA's practice of varying each node's address only within a defined IP range can thus be supported by existing infrastructure.

6.5 Conclusion

We have shown that dynamics mechanisms can introduce security benefits into the network architecture and can be implemented without significantly disrupting network operations. Using SDNA as an example, we have successfully demonstrated these techniques using real laptops and network hardware to show that this concept works in practice by providing access to authorized users while simultaneously denying service availability to attackers. Such an approach allows a legitimate user to access needed services, with the minor inconvenience of using a separate domain name and providing authentication credentials for highly sensitive services if such policies are not already in place. The attacker, however, is unable to access arbitrary services due to the use of unobservable authentication, information concealment, redirection, and various types of abstraction of addresses and names. Based on the evaluation of our prototype, we believe such dynamics significantly reduce the ability of an attacker to gather information about and subsequently spread within an enclave.

Acknowledgements The authors would like to thank AFRL for funding this research under contracts FA8750-10-C-0089 and FA8750-11-C-0179. We would like to thank our program manager Mr. Walt Tirenin from AFRL and Mr. Lynn Meredith from Lockheed Martin for their valuable suggestions and advice during this project.

Note

Approved for Public Release; Distribution Unlimited: 88ABW-2011-#1811, 29 MAR 2012

References

1. S. M. Bellovin, A. Keromytis, and B. Cheswick, "Worm propagation strategies in an IPv6 Internet," *;login:*, pp. 70–76, February 2006.
2. Panda Security, "2nd international barometer of security in smbs," Report, July 2010. [Online]. Available: http://press.pandasecurity.com/wp-content/uploads/2010/08/2nd-International-Security-Barometer.pdf
3. W. J. Lynn, "Defending a new domain," *Foreign Affairs*, vol. 5, no. 89, September/October 2010.
4. P. Dasgupta, C. K. S., and S. K. Gupta, "Vulnerabilities of PKI based smartcards," in *Proc. of IEEE Military Communications Conference (MILCOM)*, Orlando, FL, USA, October 2007.
5. McAfee, "Unified secure access solution for network access control," Datasheet. [Online]. Available: http://www.mcafee.com/us/local_content/datasheets/ds_nac.pdf
6. J. Yackoski, P. Xie, H. Bullen, J. Li, and K. Sun, "A self-shielding dynamic network architecture," in *MILCOM*, Baltimore, MD, USA, November 2011.
7. T. D. Morgan, "IPv6 address cookies: Mitigating spoofed attacks in the next generation internet," Master's thesis, Northwestern University, 2006.
8. T. Narten, G. Huston, and L. Roberts, "IPv6 Address Assignment to End Sites," RFC 6177 (Best Current Practice), Internet Engineering Task Force, Mar. 2011.
9. S. Kent and K. Seo, "Security Architecture for the Internet Protocol," RFC 4301 (Proposed Standard), Internet Engineering Task Force, Dec. 2005.
10. A. Lenstra and E. Verheul, "Selecting cryptographic key size," *Cryptography*, vol. 14, no. 4, pp. 255–293, 2001.
11. Cisco Systems, Inc., "Cisco express forwarding," Whitepaper, 1997. [Online]. Available: http://packetstormsecurity.org/defcon10/MoreInfo/CiscoExpressForwardingCEF.pdf

Chapter 7
Moving Target Defenses in the Helix Self-Regenerative Architecture

Claire Le Goues, Anh Nguyen-Tuong, Hao Chen, Jack W. Davidson,
Stephanie Forrest, Jason D. Hiser, John C. Knight, and Matthew Van Gundy

Abstract In this chapter we describe the design, development and application of the Helix Metamorphic Shield (HMS). The HMS: (1) continuously shifts the program's attack surface in both the spatial and temporal dimensions, and (2), reduces the program's attack surface by applying novel evolutionary algorithms to automatically repair vulnerabilities. The symbiotic interplay between shifting and reducing the attack surface results in the automated evolution of new program variants whose quality improves over time.

7.1 Introduction

Despite years of research warning of the dangers of the software monoculture, most systems today are still deployed in a relatively static configuration. An attack that works on one system is easily and quickly adapted to work on all similarly-configured systems. Even when software vendors regularly release security-critical patches, the window of vulnerability remains unacceptably high. Patches are not released quickly enough to combat zero day-attacks–attacks that take advantage of latent vulnerabilities known to attackers (but not necessarily known to defenders). Further, even when such patches are available, they are often not applied in a timely

C. Le Goues (✉) • A. Nguyen-Tuong • J.W. Davidson • J.D. Hiser • J.C. Knight
Intelligent Automation, Inc., 15400 Calhoun Drive, Suite 400, Rockville, MD 20855, USA
e-mail: legoues@cs.virginia.edu; nguyen@cs.virginia.edu; jwd@cs.virginia.edu;
hiser@cs.virginia.edu; knight@cs.virginia.edu

H. Chen • M. Van Gundy
University of California, Davis, 1 Shields Avenue Davis, CA, USA
e-mail: mdvangundy@ucdavis.edu; hchen@ucdavis.edu

S. Forrest
University of New Mexico
e-mail: forrest@cs.unm.edu

S. Jajodia et al. (eds.), *Moving Target Defense II: Application of Game Theory and Adversarial Modeling*, Advances in Information Security 100, DOI 10.1007/978-1-4614-5416-8_7, © Springer Science+Business Media New York 2013

manner. To remedy this situation, the notion of a moving target defense (MTD) has been put forward as a "game-changing" capability [20]. A moving target defense seeks to thwart attacks by invalidating knowledge that an adversary must possess to mount an effective attack against a vulnerable target.

In this chapter, we describe the design, development and application of the Helix metamorphic shield (HMS). The HMS: (1) continuously shifts the program's attack surface in both the spatial and temporal dimensions, and (2) reduces the program's attack surface by using novel evolutionary algorithms to automatically repair vulnerabilities. Continuously shifting the attack surface both increases the effort required for a successful attack and results in "noisier" attacks, as adversaries must repeatedly probe targeted programs to reveal critical information, e.g., a random key. Helix then turns the table on adversaries and uses information contained in these probing attempts as hints for automatically generating, vetting and deploying candidate patches. Furthermore, the presence of the HMS sets up an interesting dynamic between competing adversaries. A single vulnerability can lead to attacks of varying severity depending on the attack payload and the value of the target. An adversary that launches a low-value attack that results in a Helix repair becomes a spoiler for other adversaries eyeing higher-value targets.

The HMS is under active development. In its current form, however, the HMS already displays the following characteristics:

- The HMS has demonstrated the ability to continuously shift the attack surface of programs, thereby presenting adversaries with an ever-changing attack surface.
- The HMS concept applies at various levels of the software stack, ranging from high-level web applications to executable binaries.
- The HMS generates repairs for both security-critical and non-security critical vulnerabilities. We have demonstrated the automated generation and vetting of patches for self-repair to provide protection against a wide-range of attack classes, including infinite loops, segmentation faults, remote heap buffer overflows, non-overflow denials of service, local stack buffer overflows, and format string vulnerabilities, among others.
- The HMS turns the table on adversaries. Helix uses information revealed by adversaries to automatically repair programs.
- The HMS has been demonstrated in a closed-loop system, repairing programs automatically in response to Helix attack sensors.
- The HMS leverages inexpensive cloud computing infrastructures for its analysis engine. We have been able to repair real-world programs using Amazon's EC2 cloud infrastructure for less than $8 per bug.

7.1.1 Helix Architecture

Figure 7.1 provides a high-level overview of the Helix Metamorphic Shield architecture. Helix takes as input software in source or binary form and performs the necessary transformations to augment the input software with the ability to

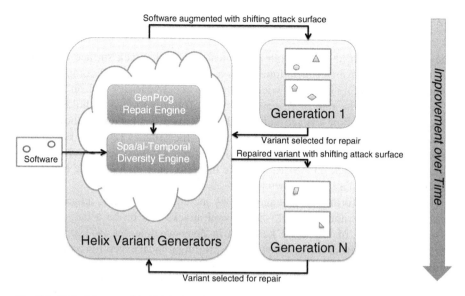

Fig. 7.1 Helix Metamorphic Shield Architecture. Every software generation is augmented with a continuously shifting attack surface using spatial-temporal diversity transformations. Over time, via its GenProg component, Helix automatically reduces the attack surface by automatically generating and vetting candidate repair patches

shift its attack surface on a continuous basis. The first generation of the deployed software retains the original vulnerabilities (shown as two circles in the box labelled Software on the left-hand side in the figure) but is already able to increase the attackers' workload for mounting a successful attack through the use of both spatial and temporal diversity techniques (shifting attack surface shown as vulnerabilities with various shapes). Section 7.2 describes how the spatio-temporal diversity engine imbues software with a dynamically shifting attack surface, and discusses security and performance results for both binary executables (Sect. 7.2.1) and web applications (Sect. 7.2.2).

When attack attempts are thwarted and detected, Helix seeks to automatically patch the targeted vulnerabilities via the GenProg Repair Engine (Sect. 7.3). GenProg uses evolutionary algorithms to create and vet candidate repair patches. The repair search and validation process is computationally expensive but is inherently parallelizable. In Sect. 7.3, we present results for automatically repairing real-world programs using the Amazon EC2 cloud infrastructure. Once a repaired variant (or set of variants) is generated and selected for deployment, Helix again will augment the variant with the ability to continuously shift its attack surface. The net result of this process is the creation of software variants that improve over time as each new generation contains fewer vulnerabilities (Generation N in Fig. 7.1).

In the scenario just outlined, the repair process was triggered reactively as a result of detecting potential attacks. Section 7.4 presents future work opportunities such as triggering repair proactively based on a variety of possible events.

7.2 Continuously Shifting the Attack Surface

The primary insight underlying our dynamically shifting attack surface is a combination of static diversity techniques with a fast-moving temporal component. A natural, but misleading intuition, would be that the effectiveness of such an approach is proportional to the rate of re-randomization. This intuition only holds true in the case of information leakage vulnerabilities, in which attackers probe a target system to expose or infer knowledge used in further attacks [20, 33]. For brute-force attacks in which attackers exhaustively perform a state-space search, such as the derandoming attack on ASLR, dynamic diversity only increases the attacker's workload by at most a factor of two [46].

Examples of information leakage attacks include side-channel attacks [8, 9], format string vulnerabilities [12], incremental attacks that probe a target system to infer knowledge of the secret key used in diversity defenses [47], or careless errors that simply reveal secret keys (such as printing it in an exception handler). To study the effectiveness of dynamically shifting the attack surface to protect against information leakage attacks, we developed an analytic model that studies the effect of re-randomization as it relates to a given rate of information leakage [33]. The model validates the notion that the rate of re-randomization should be faster than the rate at which an attacker can infer and use information about a target system.

For network-facing servers, the implication is that the attack surface must be shifted at a very high frequency, since attack probes only take a few milliseconds. In the next subsections, we show how the Helix architecture meets this requirement. First, we describe results in designing and prototyping a dynamic version of instruction-set randomization (ISR) that re-randomizes binaries at a rate of every 100 ms. We then show our results for another dynamic variant of ISR at the web-application level, potentially operating at the rate of every network request.

7.2.1 Dynamic Diversity for Protecting Binary Executables

In designing techniques for protecting binaries via dynamic diversity, we put forth the following set of requirements:

- The technique should operate on x86 binaries directly.
- The technique should proactively and continuously shift the attack surface.
- The technique should operate efficiently and at a rate that is fast enough to provide protection.
- The software architecture should be flexible and allow for a wide variety of possible diversity (and non-diversity) transformations.

These requirements were motivated by the need for more effective protection against information leakage attacks and the desire for the metamorphic shield to be practical (i.e., efficient and easily deployable), and effective as a generic platform for deploying defenses on arbitrary binaries.

Fig. 7.2 Strata
Application-Level Virtual
Machine. Any step in Strata's
fetch-decode-translate-
execute cycle is easily
replaced or augmented with
new functionality. Helix
leverages Strata's flexibility
to implement a wide variety
of security transformations
directly on executable
binaries

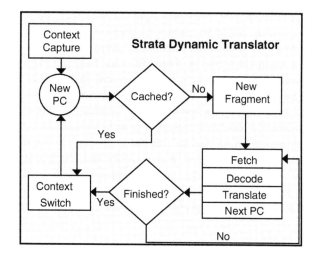

7.2.1.1 Dynamic Binary Rewriting: Strata Virtual Machine

To fulfill these requirements, we based our architecture on software dynamic translation (SDT) techniques. SDT enables software malleability and adaptivity at the instruction level, by providing facilities for transparent binary run-time monitoring and modification. SDT can affect an executing program by injecting new code, modifying existing code, or arbitrarily controlling program execution.

Examples of software dynamic translation systems include Strata [42–44], Pin [29], HDTrans [48] and DynamoRIO [25]. The flexibility afforded by software dynamic translations makes them well-suited for implementing a wide range of security transformations and policies. SDTs have been used in numerous security applications, including:

- Augmenting binaries with arbitrary sensors and actuators [11].
- Restricting control flow transfers [25].
- Thwarting return-oriented programming attacks by monitoring suspicious code sequences [10] or by relocating instructions [16].
- Diversity techniques [16, 18, 33, 35, 38, 39, 55].
- Fine-grained applications of access-control policies [36].
- Regulating resource consumptions [43].
- Protecting against fine-grained memory errors [15].

Many of these techniques have been incorporated and developed within the Helix project [11, 16, 33, 38]. Within the context of the moving-target defense, we apply Strata to dynamically shift the attack surface. We use a dynamic version of instruction-set randomization (ISR) as an exemplar. Strata was designed to be easily reconfigured and retargeted for new applications and computing platforms.

As shown in Fig. 7.2, Strata is organized as a virtual machine that mediates execution of an applications instructions. Strata dynamically loads a binary application and mediates application execution by examining and possibly translating an application's instructions before they execute on the host CPU. Translated application instructions are held in a Strata-managed code cache called the fragment cache. A fragment is the basic unit of translation, similar to a basic block. Once a fragment finishes execution, the Strata VM captures and saves the application context (e.g., PC, condition codes, registers, etc.). Following context capture, Strata processes the next application instruction. If a translation for this instruction has been cached, a context switch restores the application context and begins executing cached translated instructions on the host CPU. Otherwise, the instruction is translated (and, possibly, instrumentated) and the translation is placed into the fragment cache and is executed on the host CPU.

In the next section, we describe the dynamic instruction set randomization algorithm and its implementation over the Strata framework.

7.2.1.2 Dynamic Instruction Set Randomization

Harold Thimbleby first proposed using randomization to create a unique instruction set as a technique for preventing the spread of viruses [51]. Researchers at the University of New Mexico and Columbia University independently proposed ISR as a method for protecting against code-injection attacks [4, 5, 24]. Both groups originally implemented ISR prototypes for the x86 using emulation (Valgrind at New Mexico [32] and Bochs at Columbia [27]. Subsequent work used software dynamic translation to improve the efficiency of ISR [18, 35, 55].

A simple but effective implementation of ISR is to encode at load time (or earlier) the native binary form of a program using an XOR key [4, 5, 24, 33]. Just prior to execution, the program is decoded using the same XOR key to recover the original instruction stream. Injected code that is not encoded will likely result in the execution of random instructions that will lead to the target program crashing.

We developed a tool to analyze x86 ELF binary programs and to identify the ranges where executable instructions can exist. When Strata starts up, it encrypts code sections using a simple XOR scheme with an n-byte key:

$$P' = P \oplus Key$$

Strata intercepts the dynamic loading of libraries and performs a similar operation, typically using the same original key K, though different keys may be used.

To recover the original instructions, a decryption module between the fetch and decode modules of the Strata virtual machine applies the following transformation:

$$P = Key \oplus P'$$

An attack that attempts to inject code will most likely result in a program crash, as the decoding step will transform the injected code into random instructions. Existing XOR-based implementations exploit the likelihood of a program crash, and only shift the attack surface at load-time, randomizing the key on program startup. However, forking servers in which children processes are spawned to handle requests are common, and by default will inherit their parents' key.

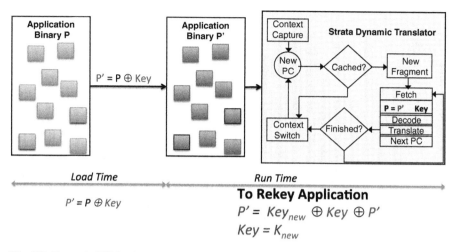

Fig. 7.3 Dynamic ISR Implementation. At load-time, the program text is encrypted using an XOR key. The Strata VM is augmented with a decryption module that reapplies the key to recover the original program text. Throughout program execution, the program text is re-encrypted under a new key resulting in a dynamically shifting attack surface

To address this concern, instead of randomizing the key only at load-time, our dynamic instruction set randomization technique continuously shifts the attack surface during program execution. Figure 7.3 illustrates our dynamic ISR implementation using the Strata virtual machine.

Rekeying the application consists of applying the old XOR key, followed by the application of a new random XOR key:

$P' = Key_{new} \oplus Key \oplus P$

$Key = Key_{new}$

Subsequently, the decryption module uses the new random key to recover the original program text.

7.2.1.3 Results

We evaluated the performance of our XOR-based ISR implementation of a Metamorphic Shield using the SPEC2000 benchmark. We present performance results for a re-randomization rate of 100 ms. All performance numbers were averaged over three runs for each of the program in SPEC2000. These numbers were obtained using version 8 of Fedora Core Linux, running in a VMWare image on a dedicated Mac Pro.

Figure 7.4 shows the performance of executing the benchmarks with and without the metamorphic shield. For a rekeying rate of 100 ms, the performance of the metamorphic shield is essentially the same as that of running the Strata virtual machine. This result is encouraging because it indicates that the metamorphic

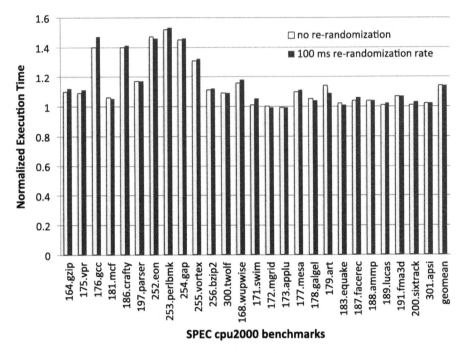

Fig. 7.4 Performance overhead of dynamic ISR over native execution. Average performance overhead over native execution is 14%. Dynamic ISR at a 100 ms re-randomization rate adds virtually no overhead to the performance of running the Strata virtual machine

shield adds virtually no overhead beyond that of Strata itself. Despite measuring the performance on an unoptimized configuration of Strata, the overall average performance overhead is only 14%.

7.2.1.4 Discussion

Sovarel et. al demonstrated incremental information-leakage attacks on weak XOR-based ISR implementations [47]. In their setup, the XOR key was not re-randomized at load-time, such as in a web server that spawns child processes to service requests. Their attack consisted of two phases. In the first phase, the XOR key is incrementally inferred via the judicious use of attack probes. By repeatedly leaking information, they were able to reconstruct the XOR key in full. Once the key is determined, the attack payload is first XOR'ed using the key. The target program will then proceed to reapply the XOR-key, executing the attacker's payload.

Is a scheme that re-randomizes the key at a 100 ms refresh rate sufficient to thwart incremental attacks? Answering this question requires making real-time assumptions about the probe rate. For example, the average probe time in the attack by Sovarel et al. is approximately 20 ms. A 100 ms rate therefore corresponds to re-randomizing after every fifth probe. However, a motivated adversary could control a botnet, issuing probes in parallel. Our analytical model showed that re-randomization should be performed frequently, as effectiveness depends critically on the re-randomization rate. The difference in the probability of attack success when re-randomizing after every 100th probe or every 4th probe spans six orders of magnitude [13, 33].

Instead of re-randomizing based on a real-time trigger, we are investigating the performance of re-randomizing based on event-driven triggers such as system calls. Since we cannot readily distinguish between normal traffic and attack probing traffic, we need to assume conservatively that every packet read over the network is potentially a probe. In the limit, we would like to re-randomize after every read system call. Furthermore, we are investigating the use of anomaly detection techniques to distinguish between normal traffic and attack probes and thereby reduce the required rate of re-randomization.

The work just described sought to continuously shift the attack surface at the binary level. We next present Noncespaces, a component of the HMS that continuously shifts the attack surface at the level of web applications.

7.2.2 Dynamic Diversity for Protecting Web Applications

Cross-site scripting (XSS) attacks pose a serious threat to the security of modern web applications. In this section, we present *Noncespaces*, an approach inspired by instruction set randomization [4, 24] that thwarts such attacks by shifting the attack surface of web applications. Noncespaces is an end-to-end mechanism that allows a server to identify untrusted content, reliably convey this information to the client, and allow the client to enforce a security policy on the untrusted content. Noncespaces randomizes (X)HTML tags and attributes to identify and defeat injected malicious web content, extending the concept of randomization up the software stack. Randomization serves two purposes. First, it identifies untrusted content so that the client can use a policy to limit the capabilities of untrusted content. Second, it prevents the untrusted content from distorting the document tree. Since the randomized tags are not guessable, the attacker cannot embed proper delimiters in the untrusted content to split the containing node without causing parsing errors.

7.2.2.1 Background

A cross-site scripting (XSS) vulnerability allows an attacker to inject malicious content into web pages served by a trusted web application. Because the browser receives the malicious content from a trusted server, the malicious content runs with the same privileges as trusted content, which allows it to run malicious code within the browser, impersonate the user to trusted servers, steal a victim user's private data and authentication credentials, or present forged content to the victim. Such attacks may be reflected or stored; in both scenarios, untrusted user input is returned to a victim user—immediately in the case of a reflected XSS attack or at some later time in the case of a stored XSS attack.

Currently, web browsers protect multiple web applications running within the same browser instance by isolating them according to the Same Origin Policy, which prevents web applications from accessing the private data of other web applications. However, this policy presumes first, that all content from a single web application is equally trustworthy, and second, that all content can be granted access to all data associated with the application. To prevent these vulnerabilities, all the untrusted (user-contributed) content in a web page must be sanitized. However, proper sanitization is very challenging. The context in which untrusted data is interpreted determines the forms of sanitization that are appropriate. There are many ways for an attacker to take advantage of the discrepancy between the way sanitization is performed by the server and the way the browser interprets the content [40]. Alternatively, one could let the client sanitize untrusted content. However, without the server's help, the client cannot distinguish between trusted and untrusted content in a web page, since both appear to originate from the trusted server.

We can avoid ambiguity between the client and server by requiring the server to identify untrusted content and requiring the client to ensure that it is displayed safely. However, challenges remain. After the server identifies untrusted content, it needs to tell the client the locations of the untrusted content in the document tree. However, the untrusted content can evade sanitization by distorting the document tree (without executing). To achieve this, the untrusted content can contain node delimiters that split the original node, where untrusted content resides, into multiple nodes. This is known as a *Node-splitting attack* [21]. To defend against this attack without restricting the richness of user provided content, the server must take care to remove only those node delimiters which would introduce new trusted nodes. Noncespaces addresses these concerns by randomizing the XHTML namespace, allowing the client to enforce security policies that limit the capabilities of untrusted content and prevent untrusted content from distorting the document tree.

7.2.2.2 Approach

The goal of Noncespaces is to allow the client to safely display documents that contain both trusted content generated by a web application and untrusted user-provided content. The browser enforces a configurable security policy to eliminate the client-server semantic gap and to adapt to differing security needs. Such a policy specifies the browser capabilities that each type of content can exercise, thus restricting the capabilities of attacker-injected malicious content.

The client must be able to determine the trustworthiness of all content in a document to faithfully enforce such a server-specified policy. Therefore, the server must first classify content into discrete trust classes. The server then must

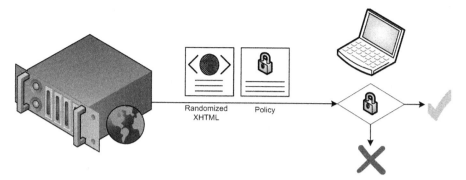

Fig. 7.5 Noncespaces Overview. The server delivers a XHTML document with randomized namespace prefixes and a policy to the client. The client accepts the document only if it is a well-formed XML document and satisfies the policy

```
1  <!DOCTYPE html PUBLIC "-//W3C//DTD XHTML 1.1//EN"
2      "http://www.w3.org/TR/xhtml11/DTD/xhtml11.dtd">
3  <html xmlns="http://www.w3.org/1999/xhtml" lang="en">
4  <head>  <title>nile.com : ++Shopping</title>  </head>
5  <body>  <h1 id="title">{item_name}</h1>
6    <p class='review'>{review.text}
7      -- <a href='{review.contact}'>{review.author}</a>  </p>
8  </body>
9  </html>
```

Fig. 7.6 Vulnerable web page template used to render dynamic web pages

communicate the content, trust classification, and policy to the client. Finally, the client can enforce the policy. This process is depicted in Fig. 7.5.

As long as the server's content classification is conservative, the server faithfully communicates its classifications to the client, and the client faithfully enforces the server-specified policy, untrusted content will be confined to the capabilities expressly permitted to it by the policy. This ensures that XSS attacks will not succeed. *Communicating Trust Information.* The server securely communicates trust information in (X)HTML to the client using randomization. We associate a different randomization function with each content trust class. The names of all elements and attributes in a trust class are remapped according to the associated randomization function so that no injected content can correctly name (X)HTML elements or attributes in other trust classes.

Consider the vulnerable web template in Fig. 7.6. We can defeat XSS attacks against this document by annotating it. For example, let the randomly chosen string r60 denote trusted content. For HTML documents, we can prefix trusted tags and attributes with our random identifier, shown in Fig. 7.7. For XHTML documents, we can preserve the original XML semantics of the document while annotating by using our random identifier as an XML namespace prefix, shown in Fig. 7.8.

```
 1  <!DOCTYPE html PUBLIC "-//W3C//DTD XHTML 1.1//EN"
 2      "http://www.w3.org/TR/xhtml11/DTD/xhtml11.dtd">
 3  <r60:html xmlns="http://www.w3.org/1999/xhtml" r60:lang="en"
 4      xmlns:r60="http://www.w3.org/1999/xhtml">
 5  <r60:head><r60:title>nile.com : ++Shopping</r60:title></r60:head>
 6  <r60:body><r60:h1 r60:id="title">Useless Do-dad</r60:h1>
 7    <r60:p r60:class='review'></p><script>attack()</script><p>
 8      -- <r60:a href=''></r60:a>  </r60:p>
 9  </r60:body>
10  </r60:html>
```

Fig. 7.7 Random prefix applied to trusted content in an HTML document containing a node-splitting attack injected by a malicious user

```
 1  <!DOCTYPE html>
 2  <r60html r60lang="en">
 3  <r60head>  <r60title>nile.com : ++Shopping</r60title>  </r60head>
 4  <r60body>  <r60h1 r60id="title">Useless Do-dad</r60h1>
 5    <r60p r60class='review'></p><script>attack()</script><p>
 6      -- <r60a href=''></r60a>  </r60p>
 7  </r60body>
 8  </r60html>
```

Fig. 7.8 Random prefix applied to trusted content in an XHTML document containing a node-splitting attack injected by a malicious user

Attackers cannot inject malicious content and cause it to be interpreted as trusted (as in a node-splitting attack) because they do not know the random prefix. They also cannot escape from the enclosing paragraph element, because they do not know the random prefix, and therefore cannot embed a closing tag with this prefix (in the HTML document, the `<script>` element is the child of an `<r60p>` element, not a `<p>` element). In the XHTML document, a closing tag that tries to close an open tag with unmatching prefixes will lead to an XML parse error.

To prevent attackers from guessing these (namespace) prefixes, we choose them uniformly at random each time a response is rendered, hence the term Noncespaces. Given a prefix space of appropriate size, knowing the random prefixes in one instance of the document does not help attackers predict prefixes in future instances of the document.

However, naïvely prohibiting all untrusted content will not work because most modern web applications are designed to accept some amount of rich content from users. Though we can use randomization to ensure integrity of trust class information, in practice, we still need a policy that places appropriate constraints on such user-provided content. Therefore, we also provide a mechanism for the server to specify a policy for the client to enforce when rendering the document. This mechanism is provided by new HTTP protocol headers, in which such a policy is specified, and a policy language used to describe the constraints.

```
1  # Restrict untrusted content to safe subset of XHTML
2  namespace  x   http://www.w3.org/1999/xhtml
3  # Declare trust classes
4  trustclass trusted
5  trustclass untrusted
6  order untrusted < trusted
7
8  #Policy for trusted content
9  allow //x:*[ns:trust-class(., "=trusted")]  # all trusted elements
10 allow //@x:*[ns:trust-class(., "=trusted")] # all trusted attributes
11
12 # Allow safe untrusted elements
13 allow //x:b   |  //x:i   |  //x:u  |  //x:s  |  //x:pre  |  //x:q
14 allow //x:a   |  //x:img  |  //x:blockquote
15
16 # Allow HTTP protocol in the <a href> and <img src> attributes
17 allow //x:a/@href[starts-with(., "http:")]
18 allow //x:img/@src[starts-with(., "http:")]
19
20 # Deny all remaining elements and attributes
21 deny //*   |  //@*
```

Fig. 7.9 Noncespaces policy restricting untrusted content to BBCode [7]

Policy Specification. A Noncespaces policy specifies the browser capabilities that can be invoked by content in a given trust class. Figure 7.9 shows an example policy for XHTML documents. We designed the policy language to be similar to a firewall configuration language. Comments begin with an # character and extend to the end of the line. A minimal policy consists of a sequence of allow/deny rules. Each rule applies a policy decision—allow or deny—to a set of document nodes matched by an XPath expression; XPath is well-suited for this domain because it was designed to query content from hierarchical documents.

Noncespaces additionally provides basic XPath functions for string normalization and additional boolean functions for matching based on trust class or whether an attribute value has changed from the language default. The example policy in Fig. 7.9 specifies two trust classes, trusted and untrusted. There are no restrictions on which tags and attributes can appear in trusted content. Only tags and attributes that correspond to BBCode are allowed in untrusted content: stylistic markup, links to other HTTP resources, and images. Note that lines 17–18 only permits link and image tags to specify URLs for the (non-script) HTTP protocol.

When checking that a document conforms to a policy, the client considers each rule in order and matches the XPath expression against the nodes in the document's Document Object Model. When an allow rule matches a node, the client permits the node and will not consider the node when evaluating subsequent rules. When a deny rule matches a node, the client determines that the document violates the policy and will not render the document. To provide a fail-safe default, if any nodes remain unmatched after evaluating all rules, we consider those nodes to be policy violations (i.e. all policies end with an implicit deny `//*|//@*`). In the event that a policy author wishes to override the default behavior in order to specify a blacklist policy, he can specify allow `//*|//@*` as the last rule to allow all as of yet unmatched nodes.

Client Enforcement. Web browsers must ensure that a Noncespaces-encoded response conforms to the policy before rendering it. The overhead involved in policy retrieval should be minimal, given that most web pages are assembled from multiple requests. Client-side enforcement of the policy is necessary because it avoids possible semantic differences between the policy checker and the browser, which might lead the browser to interpret a document in a way that violates the policy even though the policy checker has verified the document.

7.2.2.3 Results

We evaluated Noncespaces to ensure that it is able to prevent a wide variety of XSS attacks. We tested Noncespaces against six XSS exploits targeting two vulnerable applications. The exploits were crafted to exhibit the various forms that an XSS attack may take [53]. The applications used in this evaluation were a version of TikiWiki [52] with a number of XSS vulnerabilities and Trustify, a custom web application that we developed to cover all the major XSS vectors.

We began by developing policies for each application. Because TikiWiki was developed before Noncespaces existed, it illustrates the applicability of Noncespaces to existing applications. We implemented a straightforward 37-rule, static-dynamic policy that allows unconstrained static content but restricts the capabilities of dynamic content to that of BBCode (similar to Fig. 7.9). We also had to add exceptions for trusted content that TikiWiki generates dynamically by design, such as names and values of form elements, certain JavaScript links implementing collapsible menus, and custom style sheets based on user preferences.

For Trustify, our custom web application, we implemented a policy that does not take advantage of the static-dynamic model. Instead, the policy takes advantage of Noncespaces's ability to thwart node splitting attacks to implement an ancestry-based sandbox policy similar to the noexecute policy described in BEEP [21]. This policy denies common script-invoking tags and attributes from any namespace (e.g., `<script>` and `onclick`) that are descendants of a `<div>` tag with the class=``sandbox'' attribute. (Note: the policy does not attempt to be exhaustive. It does not enumerate non-standard browser-specific tags and attributes.) To allow the rules to apply to elements and attributes in any namespace we use the common XPath idiom of matching by each node's `local-name()`. An excerpt of the 28 line policy is given in Fig. 7.10.

We first verified that each exploit succeeded without Noncespaces. We then enabled Noncespaces and verified that all exploits were blocked as policy violations.

```
1  trustclass unclassified
2
3  # Blacklist (possibly incomplete)
4  deny //*[local-name() = 'div' and @*[local-name() = 'class' \
5         and . = 'sandbox']]\
6      //*[local-name() = 'script']
7  deny //*[local-name() = 'div' and @*[local-name() = 'class' \
8         and . = 'sandbox']]\
9      //@*[local-name() = 'onload' \
10         or local-name() = 'onunload'\
11         or local-name() = 'onclick' \
12         or local-name() = 'onmousedown' \
13         or local-name() = 'onmouseover' \
14         or local-name() = 'onfocus' \
15         or local-name() = 'onsubmit' \
16         or (local-name() = 'src' \
17            and starts-with(ns:tolower(normalize-space(.)), \
18                            "javascript:"))]
19 # Allow everything else
20 allow //*
21 allow //@*
22 allow //namespace::*
```

Fig. 7.10 Excerpt from an ancestry-based sandbox policy that denies all potential script-invoking tags and attributes that are descendants of a <div> node with the class=''sandbox'' attribute

7.2.2.4   Performance Evaluation

Our performance evaluation seeks to measure the overhead of Noncespaces in terms of response latency and server throughput. Our test infrastructure consisted of the TikiWiki application that we used for our security evaluation running in a VMware virtual machine with 512 MB RAM running Fedora Core 3, Apache 2.0.52, and mod_php 5.2.6. The virtual machine ran on an Intel Pentium 4 3.2 GHz machine with 1 GB RAM running Ubuntu 7.10. Our client machine was an Intel Pentium 4 2 GHz machine with 256 MB RAM running Ubuntu 8.10 Server. These results represent an upper bound on performance penalty as we have spent no effort optimizing our Noncespaces prototype. In each test we used ab [1] to retrieve an application page 1,000 times. We varied the number of concurrent requests between 1, 5, 10, and 15, and the configuration of the client and server between the following:

- Baseline: measures original web application performance before applying Noncespaces.
- Randomization Only: measures impact of Noncespaces randomization on server without policy validation on client-side.
- Full Enforcement: measures the end-to-end impact of Noncespaces.

We ran three trials with each test configuration against the TikiWiki application. The response latency shows that enabling Noncespaces randomization on the server increased response time by at most 14%. Enabling the policy checking proxy resulted in response times that were at most 32% higher than the baseline response time. Though the overhead may appear significant at first glance, during interactive use latency typically increased by no more than 0.6 s.

We also examine the effect of Noncespaces on server throughput. With randomization enabled throughput is reduced by about 10%. After enabling policy checking, the throughput decreases by an additional 3% for higher numbers of concurrent requests. Because policy checking is performed on the client side, multiple simultaneous client requests has a minimal effect on server throughput.

7.2.3 Summary: Continuously Shifting the Attack Surface

We have demonstrated the feasibility of quickly shifting a program's attack surface using dynamic variations of instruction-set randomization for both binaries and web applications. Based on this experience, we believe that it is feasible to continuously shift the attack surface at all levels of the software stack, and using a variety of possible diversity techniques. Preliminary performance results show that shifting the attack surface continuously can be performed at reasonable cost.

Regardless of how quickly the attack surface is shifted, diversity techniques do not fundamentally address the underlying vulnerability that enables the attack. For example, unsuccessful attempts to inject binary code against an ISR-protected server will typically result in a program crash. Similarly, attempts to subvert other diversity techniques such as address-space layout randomization, will also result in a program crash, allowing adversaries to launch denial-of-service attacks.

Shifting the attack surface dynamically is an arms race. Shifting must be done quickly, to stave off an adversary's efforts to both learn critical information and to launch an attack on a vulnerable program based on this knowledge. The Helix metamorphic shield (HMS) seeks to exploit this arms race in order to end it. The primary reason that attacks succeed is that adversaries have successfully discovered vulnerabilities that remain unknown to the original developers (otherwise they would, or should, have fixed them before releasing their programs). Attacks attempted against a Helix-protected program that result in a detectable event such as a program crash reveals crucial information about latent vulnerabilities. Helix turns the table on adversaries and uses such information to trigger an automated repair process. The next section discusses GenProg, a new technique based on evolutionary methods to reduce the attack surface via automated program repair.

7.3 Reducing the Attack Surface: Genetic Programming for Automatic Program Repair

Mature software projects are forced to ship with both known and unknown bugs [28], because the number of outstanding software defects typically exceeds the resources available to address them [3]. Software maintenance, of which bug repair is a major component [37], is time-consuming and expensive, accounting for as much as 90% of the cost of a software project [45] at a total cost of up to $70 billion per year in the US [23, 49]. Even security-critical bugs take an average of 28 days for developers to address [50], further imbalancing the arena in favor of the attacker.

In this section, we describe GenProg, the repair component of the HMS. GenProg is an evolutionary computation technique that efficiently and automatically repairs bugs in off-the-shelf legacy programs. Efficient repair of existing vulnerabilities allows the HMS to *reduce* the program attack surface over time. We show that GenProg can repair a large number and variety of bugs in real-world, off-the-shelf C programs. We then demonstrate its promise in a closed-loop system for automated program repair

7.3.1 Approach

GenProg is an evolutionary algorithm that repairs existing programs by selectively searching through the space of related program variants until it discovers one that avoids a known defect but retains key functionality. GenProg takes as input a program and a test suite. The program currently passes the *positive test cases*, which encode required functionality. The *negative test cases* characterize the fault under repair; the input program fails these test cases. The goal of the search is to find a variant of the input program that passes the negative test cases while continuing to pass all of the positive test cases.

Figure 7.11 shows a high-level view of GenProg architecture. GenProg conducts the search using *Genetic programming* (GP), a computational method inspired by biological evolution which evolves computer programs tailored to a particular task [26]. The GP maintains a population of program variants, or *individuals*, each of which is a candidate solution to the problem at hand. In our case, each individual corresponds to a program that varies slightly from the original program. Each individual's suitability is evaluated using a task-specific *fitness function*, and the individuals with highest fitnesses are *selected* for continued evolution. Computational analogs of biological mutation and crossover produce variations of the high-fitness programs, and the process iterates. The search terminates either when it finds a candidate solution that passes all its positive and negative test cases, or when it exceeds a preset number of iterations.

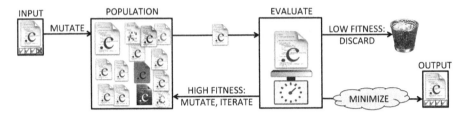

Fig. 7.11 GP architecture diagram

A significant impediment for an evolutionary algorithm like GP is the potentially infinite-size search space of potential programs. To address this, we use novel GP representations, and make assumptions about the probable nature and location of the necessary repair, improving search efficiency. The rest of this subsection provides further algorithmic details.

7.3.1.1 Representation

We have primarily applied GenProg at the source level of C programs, though it is also possible to repair programs at the ASM or ELF levels [41]. Programs can be represented at multiple levels of abstraction or granularity. For example, C programs contain both *statements*, such as the conditional statement "if (!p) \{ x=0; \}" and *expressions*, such as "0" or "(!p)". For scalability, we treat the statement as the basic unit, or gene. Thus, we never modify "(!p)" because doing so would involve changing an expression. Instead, we might delete the entire "if ..." statement, including its then- and else-branches. Additionally, individual variants do not need to store the entire program. Instead, at the source level, each variant is a *patch*, represented as sequence of edit operations. This increases scalability by avoiding the storage of redundant copies of untouched nodes.

7.3.1.2 Genetic Operators

The *mutation* and *crossover* operators produce new program variants by modifying individuals and recombining them, respectively. Because the basic unit of our representation is the statement, mutation is more complicated than the simple bit flip used in other evolutionary algorithms. A statement selected for mutation is randomly subjected to either *deletion* (the entire statement and all its sub-statements are deleted), *insertion* (another statement is inserted after it), or *swap* (two statements are replaced with one another). Crossover exchanges randomly-chosen subtrees between two individuals, which allows the GP to combine partial solutions.

7.3.1.3 Localization

We assume that software defects are local and that fixing one does not require changing the entire program. We therefore narrow the search space by biasing modifications towards statement nodes that are more closely associated with the fault [22]. Statements that are only executed by the negative test cases are weighted much more highly than those executed by both the negative and the positive test cases; statements that are not visited by the negative test case at all are not considered for mutation. In this respect, GenProg exploits information provided by an attacker.

We use the term *fix localization* to refer to the *source* of inserted or swapped code. We restrict inserted code to that which includes variables that are in-scope at the destination (so the result compiles; this presented a non-trivial concern in previous work [54]) and that are visited by at least one test case.

7.3.1.4 Fitness Function

The *fitness* function guides the search. The fitness of an individual in a program repair task should assess how well the program avoids the bug while still doing "everything else it is supposed to do." We use test cases, such as those that often ship with existing software, to measure fitness. Such testing accounts for as much as 45% of total software lifecycle costs [34], and finding test cases to cover all parts of the program and all required behavior is a difficult but well-studied problem in the field of software engineering.

The fitness function applies the edits associated with a variant to the input program and compiles the result into an executable. This executable is run against the set of positive and negative test cases, returning the weighted sum of the test cases passed. Programs that do not compile have fitness zero.

7.3.1.5 Minimizing the Repair

The first variant that passes all positive and negative test cases is called the *primary repair*. However, GP may introduce irrelevant changes on the way to a repaired variant [14]. The minimization step uses tree-structured differencing [2] to express the primary repair as a set of changes to the original program. It then uses *delta debugging* [56] to efficiently compute a subset of these changes such that the changed program passes all test cases, but dropping any additional elements causes the program to fail at least one test case. Delta debugging is conceptually similar to binary search, but it returns a set instead of a single number. We call this smaller set of changes the *final repair*; in our experiments, the final repair is typically at least an order-of-magnitude smaller than the primary repair.

7.3.2 Efficacy

In the context of an MTD system, GenProg reduces the attack surface by *improving* software in response to detected defects. This section presents results supporting our claim that GenProg provides a scalable and general approach to automatically repairing detected bugs in programs.

7.3.2.1 Scalability

Success Metrics. Bug repair has become such a pressing problem that many companies have begun offering *bug bounties* to outside developers, paying for candidate repairs. Well-known companies such as Mozilla[1] and Google[2] offer significant rewards for security fixes, with bounties raising to thousands of dollars in "bidding wars."[3] Although security bugs command the highest prices, more wide-ranging bounties are available for bugs ranging from cosmetic concerns to security vulnerabilities (e.g., those provided by Tarsnap.com[4]). These examples suggest that relevant success metrics for automatic bug repair include the fraction of queries that produce code patches, monetary cost, and wall-clock time cost. This section uses these metrics to evaluate GenProg's scalability .

Benchmarks. Benchmarks appear in the left-hand column of Table 7.1. We sought to define this set in as unbiased a manner as possible. At a high level, we selected these benchmarks by searching various program repositories to identify acceptable candidate programs (e.g., consisting of at least 50,000 lines of C code, 10 viable test cases, and 300 versions in a revision control system) and reproducible bugs within those programs. We searched systematically through each program's source history, looking for revisions that caused the program to pass test cases that it failed in a previous revision. Such a scenario corresponds to a human-written repair for the bug corresponding to the failing test case. This approach ensures that benchmark bugs are important enough to merit a human fix and to affect the program's test suite. The ultimate benchmark set consists of 8 subject C programs covering a variety of uses, comprising 5.1 MLOC and more than 10,000 test cases.

Cloud Computing Framework. *Cloud computing*, in which virtualized processing power is purchased cheaply and on-demand, is becoming commonplace and less expensive over time [31]. To evaluate the cost of repairing a bug with GenProg, we used Amazon's EC2 cloud computing infrastructure for the experiments. Each trial was given a "high-cpu medium (c1.medium) instance" with two cores and 1.7 GB

---

[1] http://www.mozilla.org/security/bug-bounty.html $3,000/bug

[2] http://blog.chromium.org/2010/01/encouraging-more-chromium-security.html $500/bug

[3] http://www.computerworld.com/s/article/9179538/Google_calls_raises_Mozilla_s_bug_bounty_for_Chrome_flaws

[4] http://www.tarsnap.com/bugbounty.html

of memory.[5] Simplifying a few details, the virtualization can be purchased as *spot instances* at \$0.074 per hour but with a one hour start time lag, or as *on-demand instances* at \$0.184 per hour.[6]

Experimental Parameters. We ran 10 random GenProg trials per bug. Each trial was terminated after 10 generations, 12 h, or when another search found a repair, whichever came first. Population size is 40; each individual was mutated exactly once per generation; and 50% of the population is retained (with mutation) on each generation (known as elitism).

Repair Results. The right side of Table 7.1 reports results. We report costs in terms of monetary cost and wall clock time from the start of the request to the final result, recalling that the process terminates as soon as one parallel search finds a repair. Results are reported for cloud computing spot instances, and thus include a one-hour start lag but lower CPU-hour costs.

GenProg repaired 55 of the defects (52%) within the allocated time/generation limits, including at least one defect for each subject program. The successful repairs return a result in 1.6 h each, on average. The 50 unsuccessful repairs required 11.22 h each, on average. The total cost for all 105 attempted repairs is \$403, or \$7.32 per successful run. These costs could be traded off in various ways. For example, an organization that valued speed over monetary cost could use on-demand cloud instances, reducing the average time per repair by 60–36 min, but increasing the average cost per successful run from \$7.32 to \$18.30.

Diverse solutions to the same problem may provide several options to developers, or enable consideration of multiple diverse attack surfaces. To investigate GenProg's utility in generating multiple repairs, we additionally allowed all of the bug trials to run to completion (instead of terminating when any trial found a repair). The "Patches per Bug" column in Table 7.1 shows how many different patches were discovered in this use-case. GenProg produced 318 unique patches for 55 repairs, or an average of 5.8 distinct patches per repaired bug. The unique patches are typically similar, often involving different formulations of guards for inserted blocks or different computations of required values. Such diverse patches can contribute to multiple semantically-equivalent variants, helping reduce the software monoculture.

7.3.2.2 Generalizability

The previous section showed that GenProg can scale to real bugs in millions of lines of real code. In this section, we substantiate our claim that GenProg provides a *general* means for program repair by evaluating it on a benchmark set designed to cover a variety of bug types.

---

[5]http://aws.amazon.com/ec2/instance-types/

[6]These August–September 2011 prices summarize CPU, storage and I/O charges; http://aws.amazon.com/ec2/pricing/

Table 7.1 Subject C programs, test suites and historical defects: defects are defined as test case failures fixed by developers in previous versions. Fifty-five of the 105 defects (52%) were repaired successfully and are reported under the "Cost per Repair" columns. The remaining 50 are reported under the "Non-Repair" columns. "Hours" columns report the wall-clock time between the submission of the repair request and the response, including cloud-computing spot instance delays. "US$" columns reports the total cost of cloud-computing CPU time and I/O. The total cost of generating these results was $403. "Patches per bug" shows the number of unique patches per bug

Program	LOC	Description	Tests	Defects repaired	Patches per Bug	Cost per non-repair		Cost per repair	
						Hours	US$	Hours	US$
fbc	97,000	Legacy programming	773	1/3	1.0	8.52	5.56	6.52	4.08
gmp	145,000	Multiple precision math	146	1/2	2.0	9.93	6.61	1.60	0.44
gzip	491,000	Data compression	12	1/5	8.0	5.11	3.04	1.41	0.30
libtiff	77,000	Image manipulation	78	17/24	6.8	7.81	5.04	1.05	0.04
lighttpd	62,000	Web server	295	5/9	4.6	10.79	7.25	1.34	0.25
php	1,046,000	Web programming	8,471	28/44	5.6	13.00	8.80	1.84	0.62
python	407,000	General programming	355	1/11	5.0	13.00	8.80	1.22	0.16
wireshark	2,814,000	Network packet analyzer	63	1/7	7.0	13.00	8.80	1.23	0.17
total or **avg**	*5,139,000*		*10,193*	*55/105*	**5.8**	**11.22 h**		**1.60 h**	

Programs and Defects. The benchmarks for these experiments are shown in the left-hand side of Table 7.2. zune is a fragment of code that caused all Microsoft Zune media players to freeze on December 31, 2008. The Unix utilities were taken from Miller et al.'s work on *fuzz testing*, in which programs crash when given random inputs [30]. The remaining benchmarks are taken from public vulnerability reports. The defects considered cover eight defect classes: infinite loop, segmentation fault, remote heap buffer overflow to inject code, remote heap buffer overflow to overwrite variables, non-overflow denial of service, local stack buffer overflow, integer overflow, and format string vulnerability.

Test Cases. For each program, we used a single negative test case that elicits the given fault. We selected a small number (e.g., 2–6) of positive test cases per program. In some cases, we used non-crashing fuzz inputs; in others, we manually created simple cases, focusing on testing relevant program functionality; for openldap, we used part of its test suite.

Experimental Parameters. We ran 100 random GenProg per each bug. Otherwise, the parameter set is the same here as in the previous subsection.

Repair Results. Table 7.2 summarizes repair results for the fifteen C programs. The "Time" column reports the average wall-clock time per trial that produced a primary repair. It does not include the minimization time, which is considerably less than the time taken to repair. Repairs are found in 357 s on average. The "Success" column gives the fraction of trials that were successful. On average, over 77% of the trials produced a repair, although most of the benchmarks either succeeded very frequently or very rarely. Low success rates can be mitigated by running multiple independent trials in parallel. The "Size" column lists the size of the final (minimized) repair diff in lines. The final minimized patch is quite manageable, averaging 5.1 lines. The "Effect" column shows a summary of the effect of the final repair, as judged by manual inspection.

Patch Effect. Manual inspection suggests that the majority of produced patches are acceptable, meaningfully changing the program semantics to guard against the error in question while otherwise maintaining functionality. Of the fifteen patches, six insert code (zune, look−u, look−s, units, ccrypt, and indent) seven delete code (uniq, deroff, openldap, lighttpd, flex, atris, and php), and two both insert and delete code (nullhttpd and wu−ftpd).

Patches that delete code do not necessarily degrade functionality: the deleted code may have been included erroneously, or the patch may compensate for the deletion with an insertion. The uniq, deroff, and flex patches delete erroneous code and do not degrade untested functionality. The openldap patch removes unnecessary faulty code (handling of multi-byte BER tags, when only 30 tags are used), and thus does not degrade functionality in practice. The nullhttpd and wu−ftpd patches delete faulty code and replace them with non-faulty code found elsewhere. The effect of the lighttpd patch is machine-specific: it may reduce functionality on certain inputs, though in our experiments, it did not.

Table 7.2 Experimental results on bugs from programs totaling 1.25M lines of source code. Size of programs given in lines of code (LOC). A † indicates an openly-available exploit. We report averages for 100 random trials. "Time" gives the average time taken for each successful trial and "Success" (how many of the random trials resulted in a repair). "Size" reports the average Unix diff size between the original source and the final repair, in lines. "Effect" describes the operations performed by an indicative final patch: a patch may insert code (I), delete code (D), or both insert and delete code (B).

Program	Lines of Code	Description	Fault	Time (s)	Success (%)	Size	Effect
zune	28	Example [6]	Infinite loop†	42	72	3	I
uniq utx	1146	Text processing	Segmentation fault	34	100	4	D
look utx	1169	Dictionary lookup	Segmentation fault	45	99	11	I
look svr	1363	Dictionary lookup	Infinite loop	55	100	3	I
units svr	1504	Metric conversion	Segmentation fault	109	7	4	I
deroff utx	2236	Text processing	Segmentation fault	131	97	3	D
nullhttpd	5575	webserver	Remote heap buffer Overflow (code)†	578	36	5	B
openldap	293k	Directory protocol	Non-overflow Denial of service†	665	100	16	D
ccrypt	7515	Encryption utility	Segmentation fault†	330	100	14	I
indent	9906	Code processing	Infinite loop	546	7	2	I
lighttpd	52k	Webserver	Remote heap buffer Overflow (variables)†	394	100	3	D
flex	19k	Lexical analyzer generator	Segmentation fault	230	5	3	D
atris	22k	Graphical game	local stack Buffer exploit†	80	82	3	D
php	764k	Scripting language	Integer overflow†	56	100	10	D
wu-ftpd	67k	FTP server	format string Vulnerability†	2256	75	5	B
Average	1,246,781			356.5	77.0	5.7	

In many cases, it is possible to insert code without negatively affecting the functionality. The zune benchmark contains an infinite loop when calculating dates involving leap years. The repair inserts code to one of three branches that decrements the day in the main body of the loop. The insertion is carefully guarded so as to apply only to relevant inputs, and thus does not negatively impact other functionality. Similar behavior is seen for look−s, where a buggy binary search over a dictionary never terminates if the input dictionary is not pre-sorted. Our repair inserts a new exit condition to the loop (i.e., a guarded break). A more complicated example is units, in which user input is read into a static buffer without bounds checks, a pointer to the result is passed to a lookup() function, and the result of lookup() is possibly deferenced. Our repair inserts code into lookup() so that it calls an existing initialization function on failure (i.e., before the return), re-initializing the static buffer and avoiding the segfault. These changes are indicative of repairs involving inserted code.

Functionality Degradation. Only one of the patches, for php, obviously degrades functionality. Disabling functionality to suppress a security violation is often a legitimate response: many systems can be operated in a "safe mode" or "read-only mode." Although acceptable in this situation, disabling functionality could have deleterious consequences in other settings. We explore this patch in more detail and evaluate its impact on functionality in the next section.

7.3.3 Closed-Loop Repair

The *automated* repair system evaluated above relies on manual initialization and dispatch of GenProg. However, automated detection techniques in the Helix system can signal the repair process to complete the automation loop. This proposed integration and the corresponding removal of the human from the loop present several areas of additional experimental concern, particularly related to the *quality* of the repairs. Incomplete test suites may lead to fragile or inadequate repairs, further compromising the system. Additionally, a closed-loop system may be subject to *false positives*, where a detector incorrectly signals the existence of a vulnerability; repairs made in response to such false positives may negatively impact the system. GenProg's success in the Helix infrastructure is partially predicated on the practical impact of its generated repairs absent human review.

This section therefore evaluates GenProg in a closed-loop repair system, with several experimental goals: (1) measure the performance impact of repair time and quality on a real, running system, including the effects of a functionality reducing repair on system throughput (2) analyze the functional quality of the generated repairs using fuzz testing and variant bug-inducing input and (3) measure the costs associated with intrusion-detection system false positives.

7.3.3.1 Closed-Loop System Overview

Our closed-loop prototype is designed with webservers in mind, because they provide compelling case studies for closed-loop automatic repair in the Helix system while allowing for evaluation on real-world programs with realistic workloads. While the webserver is run normally and exposed to untrusted inputs from the outside world, an intrusion-detection system (IDS) checks for anomalous behavior, and the system stores program state and each input while it is being processed. Our prototype system adopts an IDS that detects suspicious HTTP requests based on request features [19]. When the IDS detects an anomaly, the program is suspended, and GenProg is invoked. The negative test case is constructed from the IDS-flagged input. The positive tests consist of standard regression tests.

The efficacy of the proposed system depends on the anomaly detector's misclassification rates (false positives/negatives) and the efficacy of the repair method. The proposed system creates two new areas of particular concern. The first is the effect of an imperfect repair (e.g., one that degrades functionality not guaranteed by the positive tests) to a true vulnerability, which can potentially lead to the loss of legitimate requests or, in the worst case, new vulnerabilities. The second new concern is that a "repair" generated in response to an IDS false alarm could also degrade functionality, again losing legitimate requests. In both cases, the changed attack surface must improve the program, in that it meaningfully modifies it in the face of an attack, without introducing new vulnerabilities for attackers to exploit. The remainder of this section evaluates these concerns.

7.3.3.2 Benchmarks and Workload

We focus these experiments on three benchmarks of our benchmarks from Sect. 7.3.2 that consist of three security vulnerabilities in long-running servers: lighttpd , nullhttpd , and php; these can be seen in Table 7.2. Note that we repair the php interpreter used by an unchanging apache webserver deployment, in libphp . so. The rest of this section outlines the vulnerabilities and repairs in more detail to provide context and to illustrate our claims regarding GenProg's efficacy.

The nullhttpd webserver is a lightweight multithreaded webserver that handles static content as well as CGI scripts. Version 0.5.0 contains a heap-based buffer overflow vulnerability that allows remote attackers to execute arbitrary code using POST requests. The problem arises because nullhttpd trusts the Content−Length value provided by the user in the HTTP header of POST requests; negative values cause nullhttpd to overflow a buffer. However, there is another location in the code that similarly but correctly processes POST-data. The GenProg-generated repair changes the faulty location such that it calls the other POST-processing code, and thus correctly bounds-checks the vulnerable value. The final, minimized repair is 5 lines long. Although the repair is not the one supplied in the next release by human developers—which inserts local bounds-checking directly—it both eliminates the vulnerability and retains desired functionality.

lighttpd is a webserver optimized for high-performance environments; it is used by YouTube and Wikimedia, among others. In version 1.4.17, the fastcgi module, which improves script performance, is vulnerable to a heap buffer overflow that allows remote attackers to overwrite arbitrary CGI variables (and thus control what is executed) on the server machine. The key problem is with the fcgi_env_add function, which uses memcpy to add data to a buffer without proper bounds checks. fcgi_env_add is called many times in a loop. The repair modifies this code such that the loop exits early on very long data allocations. However, the repaired server can still report all CGI and server environment variables and serve both static and dynamic content.

php is an interpreter for a popular web-application scripting language. Version 5.2.1 is vulnerable to an integer overflow attack that allows attackers to execute arbitrary code by exploiting the way the interpreter calculates and maintains bounds on string objects in single-character string replacements. Single-character string replacement replaces every instance of a character in a string with a larger string. This functionality is implemented by php_char_to_str_ex , which handles both single-character and multi-character replacements. The repair disables the single-character case, leaving multi-character replacements untouched (multi-character replacements are not vulnerable to the attack). We use this patch to evaluate the impact of a functionality-degrading patch in the context of the closed-loop system.

Workloads. We use indicative workloads taken from the University of Virginia Computer Science Department webserver to measure program throughput pre-, during-, and post-repair. To evaluate repairs to the nullhttpd and lighttpd web-servers, we used a workload of 138,226 HTTP requests spanning 12,743 distinct client IP addresses over a 14-h period. To evaluate repairs to php, we obtained the room and resource reservation system used by the University of Virginia Computer Science Department. It totals 16,417 lines of PHP, including 28 uses of str_replace (the subject of the php repair). We also obtained 12,375 requests to this system.

We use two metrics to evaluate repair overhead and quality. The first metric is the number of successful requests a program processed before, during, and after a repair. We assume a worst-case scenario in which the same machine is used both for serving requests and repairing the program, and in which all incoming requests are dropped (i.e., not buffered) during the repair process. The second metric evaluates a program on held-out fuzz testing; comparing behavior pre- and post-repair can suggest whether a repair has introduced new errors, and whether the repair generalizes.

7.3.3.3 The Cost of Repair Time

The "Requests Lost To Repair Time" column of Table 7.3 shows the requests dropped during the repair as a fraction of the total number of successful requests served by the original program. The numbers have been normalized to the requests processed by the unmodified programs on a single day, assuming a single attack.

Table 7.3 Closed-loop repair system evaluation. Each row represents a different repair scenario and is separately normalized so that the pre-repair daily throughput is 100%. The nullhttpd and lighttpd rows show results for true repairs. The php row shows the results for a repair that degrades functionality. The False Pos. rows show the effects of repairing three intrusion detection system false positives on nullhttpd . The number after ± indicates one standard deviation. "Lost to Repair Time" indicates the fraction of the daily workload lost while the server was offline generating the repair. "Lost to Repair Quality" indicates the fraction of the daily workload lost after the repair was deployed. "Generic Fuzz Test Failures" counts the number of held-out fuzz inputs failed before and after the repair. "Exploit Failures" measures the held-out fuzz exploit tests failed before and after the repair

| | | | | Fuzz test failures | | | |
| | | | | Generic | | Exploit | |
Program	Repair made?	Requests lost to repair time	Requests lost to repair quality	Before	After	Before	After
nullhttpd	Yes	2.4% ± 0.83%	0.0% ± 0.25%	0	0	10	0
lighttpd	Yes	2.0% ± 0.37%	0.0% ± 1.53%	1,410	1,410	9	0
php	Yes	0.1% ± 0.00%	0.0% ± 0.02%	3	3	5	0
Quasi False Pos. 1	Yes	7.8% ± 0.49%	0.0% ± 2.22%	0	0	–	–
Quasi False Pos. 2	Yes	3.0% ± 0.29%	0.6% ± 3.91%	0	0	–	–
Quasi False Pos. 3	No	6.9% ± 0.09%	–	–			

Fewer than 8% of daily requests were lost while the system was offline for repairs. Buffering requests, repairing on a separate machine, or using techniques such as signature generation could reduce this overhead.

7.3.3.4   Cost of Repair Quality

The "Requests Lost to Repair Quality" column of Table 7.3 quantifies the effect of the generated repairs on program throughput. This row shows the difference in the number of requests that each benchmark could handle before and after the repair, as a percentage of total daily throughput. The repairs for nullhttpd and lighttpd do not noticeably affect their performance. Recall, however, that the php repair degrades functionality by disabling portions of the str_replace function. The php row of Table 7.3 shows that this low quality repair does not strongly affect system performance. Given the low-quality repair's potential for harm, the low "Lost" percentage for php is worth examining. Of the reservation application's 28 uses of str_replace , 11 involve replacements of multi-character substrings, such as replacing '− −' with '− −'; the repair does not affect multi-character substring replacements. Many of the other uses of str_replace occur on rare paths. For example, many uses replace underscores with spaces in a form label field. If there are no underscores in the field, the result remains correct, since the repair causes single-character str_replace to return the input. Finally, a few of the remaining uses were for SQL sanitization; because the application also uses mysql_real_escape_string , it remains safe from such attacks.

7.3.3.5   Repair Generality and Fuzzing

Two additional concerns remain. First, repairs must not introduce new flaws or vulnerabilities, even when such behavior is not tested by the input test cases. To this end, Microsoft requires that security-critical changes be subject to 100,000 fuzz inputs [17] (i.e., randomly generated structured input strings). Similarly, we used the SPIKE black-box fuzzer from immunitysec.com to generate 100,000 held-out fuzz requests using its built-in handling of the HTTP protocol. The "Generic" column in Table 7.3 shows the results of supplying these requests to each program. Each program failed no additional tests post-repair. Second, a repair must do more than merely memorize and reject the exact attack input: it must address the underlying vulnerability. To evaluate whether the repairs generalize, we used the fuzzer to generate 10 held-out variants of each exploit input. The "Exploit" column shows the results. For example, lighttpd was vulnerable to nine of the variant exploits (plus the original exploit attack), while the repaired version defeated all of them (including the original). In no case did GenProg's repairs introduce any errors that were detected by 100,000 fuzz tests, and in every case GenProg's repairs defeated variant attacks based on the same exploit, showing that the repairs were not simply fragile memorizations of the input.

7.3.3.6 Cost of Intrusion Detection False Positives

Finally, we examine the effect of IDS false positives when used as a signal to GenProg. We randomly selected three of the lowest-scoring normal requests (closest to being incorrectly labeled anomalous) and attempted to "repair" nullhttpd against them; we call these requests quasi-false positives (QFPs). The "Quasi False Pos." rows of Table 7.3 show the effect of time to repair and requests lost to repair when repairing these QFPs.

QFP #1 is a malformed HTTP request. The GenProg repair changed the error response behavior so that the response header confusingly includes HTTP/1.0 200 OK while the user-visible body retains the correct 501 Not Implemented message, but with the color-coding stripped. The header inclusion is ignored by most clients; the second change affects the user-visible error message. Neither causes the webserver to drop additional legitimate requests, as Table 7.3 demonstrates.

QFP #2 is a HEAD request; such requests are rarer than GET requests and only return header information such as last modification time. They are used by clients to determine if a cached local copy suffices. The repair changes the processing of HEAD requests so that the Cache−Control: no−store line is omitted from the response. The no−store directive instructs the browser to store a response only as long as it is necessary to display it. The repair thus allows clients to cache pages longer than might be desired. It is worth noting that the Expires: <date> also included in the response header remains unchanged and correctly set to the same value as the Date: <date> header (also indicating that the page should not be cached), so a conforming browser is unlikely to behave differently. Table 7.3 indicates request loss.

QFP #3 is a relatively standard request, whichGenProg fails to "repair". Since no repair is deployed, there is no subsequent loss to repair quality.

These experiments support the claim that GenProg repairs address given errors and without compromising functionality. It appears that the time taken to generate these repairs is reasonable and does not unduly influence real-world program performance. Finally, the danger from anomaly detection false positives is lower than that of low-quality repairs from inadequate test suites, but that both limitations are manageable. We conclude that integration of GenProg in the Helix framework viably modifies and reduces a program's attack surface in response to detected vulnerabilities, and can thus improve program performance over time.

7.4 Conclusions and Future Work

We have described the Helix metamorphic shield, which (1) continuously shifts the program's attack surface both spatially and temporally and (2), reduces the attack surface by automatically repairing existing vulnerabilities as they are detected. Taken together, these techniques allow software to change quickly enough to thwart a determined attacker and to improve over time by taking advantage of information revealed by such an attacker. Our results show that these approaches are cost-effective, applying to a wide variety of error types and at multiple layers of the software stack. We intend to continue exploring the benefits of randomization, such as by combining automatic exploit and test case generation for existing binaries to construct a closed-loop hardening system for existing binaries, enabling the proactive reduction and shifting of the program's attack surface without the need for attackers to reveal information about vulnerabilities.

Acknowledgements This research is supported by National Science Foundation (NSF) grant CNS-0716446, the Army Research Office (ARO) grant W911-10-0131, the Air Force Research Laboratory (AFRL) contract FA8650-10-C-7025, and DoD AFOSR MURI grant FA9550-07-1-0532. The views and conclusions contained herein are those of the authors and should not be interpreted as necessarily representing the official policies or endorsements, either expressed or implied, of the NSF, AFRL, ARO, DoD, or the U.S. Government.

References

1. http://httpd.apache.org/docs/2.2/programs/ab.html (2010)
2. Al-Ekram, R., Adma, A., Baysal, O.: diffX: an algorithm to detect changes in multi-version XML documents. In: Conference of the Centre for Advanced Studies on Collaborative research, pp. 1–11. IBM Press (2005)
3. Anvik, J., Hiew, L., Murphy, G.C.: Coping with an open bug repository. In: OOPSLA Workshop on Eclipse Technology eXchange, pp. 35–39 (2005)
4. Barrantes, E.G., Ackley, D.H., Forrest, S., Palmer, T.S., Stefanović, D., Zovi, D.D.: Randomized Instruction Set Emulation to Disrupt Binary Code Injection Attacks. In: Conference on Computer and Communications Security, pp. 281–289. ACM (2003)
5. Barrantes, E.G., Ackley, D.H., Forrest, S., Stefanovic, D.: Randomized instruction set emulation. ACM Transactions on Information System Security. **8**(1), 3–40 (2005). DOI http://doi.acm.org/10.1145/1053283.1053286
6. BBC News: Microsoft zune affected by 'bug'. In: http://news.bbc.co.uk/2/hi/technology/7806683.stm (2008)
7. http://www.phpbb.com/community/faq.php?mode=bbcode
8. Bernstein, D.J.: Cache-timing attacks on AES (2005). URL http://cr.yp.to/antiforgery/cachetiming-20050414.pdf
9. Brumley, D., Boneh, D.: Remote timing attacks are practical. In: Proceedings of the 12th USENIX Security Symposium, pp. 1–14 (2003)
10. Chen, P., Xiao, H., Shen, X., Yin, X., Mao, B., Xie, L.: DROP: Detecting return-oriented programming malicious code. Information Systems Security pp. 163–177 (2009)
11. Co, M., Coleman, C.L., Davidson, J.W., Ghosh, S., Hiser, J.D., Knight, J.C., Nguyen-Tuong, A.: A lightweight software control system for cyber awareness and security. Resilient Control Systems pp. 19–24 (2009)
12. Cowan, C., Barringer, M., Beattie, S., Kroah-Hartman, G.: Formatguard: Automatic protection from printf format string vulnerabilities. In: USENIX Security Symposium, (2001)
13. Evans, D., Nguyen-Tuong, A., Knight, J.C.: Effectiveness of moving target defenses. In: S. Jajodia, A.K. Ghosh, V. Swarup, C. Wang, X.S. Wang (eds.) Moving Target Defense, *Advances in Information Security*, vol. 54, pp. 29–48. Springer (2011)
14. Gustafson, S., Ekart, A., Burke, E., Kendall, G.: Problem difficulty and code growth in genetic programming. Genetic Programming and Evolvable Machines pp. 271–290 (2004)

15. Hiser, J.D., Coleman, C.L., Co, M., Davidson, J.W.: Meds: The memory error detection system. In: Symposium on Engineering Secure Software and Systems, pp. 164–179 (2009)
16. Hiser, J.D., Nguyen-Tuong, A., Co, M., Hall, M., Davidson, J.W.: ILR: Where'd my gadgets go? In: IEEE Symposium on Security and Privacy. IEEE (2012)
17. Howard, M., Lipner, S.: The Security Development Lifecycle. Microsoft Press (2006)
18. Hu, W., Hiser, J., Williams, D., Filipi, A., Davidson, J.W., Evans, D., Knight, J.C., Nguyen-Tuong, A., Rowanhill, J.: Secure and practical defense against code-injection attacks using software dynamic translation. In: Virtual Execution Environments, pp. 2–12 (2006)
19. Ingham, K.L., Somayaji, A., Burge, J., Forrest, S.: Learning DFA representations of HTTP for protecting web applications. Computer Networks **51**(5), 1239–1255 (2007)
20. Jajodia, S., Ghosh, A.K., Swarup, V., Wang, C., Wang, X.S. (eds.): Moving Target Defense - Creating Asymmetric Uncertainty for Cyber Threats, *Advances in Information Security*, vol. 54. Springer (2011)
21. Jim, T., Swamy, N., Hicks, M.: Defeating Scripting Attacks with Browser-Enforced Embedded Policies. In: International World Wide Web Conference, pp. 601–610 (2007)
22. Jones, J.A., Harrold, M.J.: Empirical evaluation of the Tarantula automatic fault-localization technique. In: Automated Software Engineering, pp. 273–282 (2005)
23. Jorgensen, M., Shepperd, M.: A systematic review of software development cost estimation studies. IEEE Transactions on Software Engineering **33**(1), 33–53 (2007)
24. Kc, G.S., Keromytis, A.D., Prevelakis, V.: Countering Code-Injection Attacks With Instruction-Set Randomization. In: Conference on Computer and Communications Security, pp. 272–280 (2003)
25. Kiriansky, V., Bruening, D., Amarasinghe, S.P.: Secure execution via program shepherding. In: USENIX Security Symposium, pp. 191–206 (2002)
26. Koza, J.R.: Genetic Programming: On the Programming of Computers by Means of Natural Selection. MIT Press (1992)
27. Lawton, K.P.: Bochs: A portable pc emulator for unix/x. Linux J. **1996**(29es), 7 (1996)
28. Liblit, B., Aiken, A., Zheng, A.X., Jordan, M.I.: Bug isolation via remote program sampling. In: Programming language design and implementation, pp. 141–154 (2003)
29. Luk, C.K., Cohn, R., Muth, R., Patil, H., Klauser, A., Lowney, G., Wallace, S., Reddi, V.J., Hazelwood, K.: Pin: Building customized program analysis tools with dynamic instrumentation. In: Programming Language Design and Implementation, pp. 190–200 (2005)
30. Miller, B.P., Fredriksen, L., So, B.: An empirical study of the reliability of UNIX utilities. Communications of the Association for Computing Machinery **33**(12), 32–44 (1990)
31. Molnar, D., Li, X.C., Wagner, D.A.: Dynamic test generation to find integer bugs in x86 binary linux programs. In: USENIX Security Symposium, pp. 67–82 (2009)
32. Nethercote, N., Seward, J.: Valgrind: a framework for heavyweight dynamic binary instrumentation. In: Programming Language Design and Implementation, pp. 89–100 (2007)
33. Nguyen-Tuong, A., Wang, A., Hiser, J., Knight, J., Davidson, J.: On the effectiveness of the metamorphic shield. In: European Conference on Software Architecture: Companion Volume, pp. 170–174 (2010)
34. Pigoski, T.M.: Practical Software Maintenance: Best Practices for Managing Your Software Investment. John Wiley & Sons, Inc. (1996)
35. Portokalidis, G., Keromytis, A.D.: Fast and practical instruction-set randomization for commodity systems. In: Annual Computer Security Applications Conference, pp. 41–48 (2010)
36. Rajkumar, R., Wang, A., Hiser, J.D., Nguyen-Tuong, A., Davidson, J.W., Knight, J.C.: Component-oriented monitoring of binaries for security. In: Hawaii International Conference on System Sciences, pp. 1–10 (2011)
37. Ramamoothy, C.V., Tsai, W.T.: Advances in software engineering. IEEE Computer **29**(10), 47–58 (1996)
38. Rodes, B.: Stack layout transformation: Towards diversity for securing binary programs. In: Doctoral Symposium, International Conference of Software Engineering (2012)

39. Rodes, B., Nguyen-Tuong, A., Knight, J., Shepherd, J., Hiser, J.D., Co, M., Davidson, J.W.: Diversification of stack layout in binary programs using dynamic binary translation. Tech. rep. (2012)
40. RSnake: XSS (Cross Site Scripting) Cheat Sheet. http://ha.ckers.org/xss.html (2008)
41. Schulte, E., Forrest, S., Weimer, W.: Automatic program repair through the evolution of assembly code. In: Automated Software Engineering, pp. 33–36 (2010)
42. Scott, K., Davidson, J.: Strata: A software dynamic translation infrastructure. In: IEEE Workshop on Binary Translation (2001)
43. Scott, K., Davidson, J.: Safe virtual execution using software dynamic translation. In: Annual Computer Security Applications Conference (2002)
44. Scott, K., Kumar, N., Velusamy, S., Childers, B.R., Davidson, J.W., Soffa, M.L.: Retargetable and reconfigurable software dynamic translation. In: International Symposium on Code Generation and Optimization, pp. 36–47 (2003)
45. Seacord, R.C., Plakosh, D., Lewis, G.A.: Modernizing Legacy Systems: Software Technologies, Engineering Process and Business Practices. Addison-Wesley Longman Publishing Co., Inc. (2003)
46. Shacham, H., Page, M., Pfaff, B., Goh, E., Modadugu, N., Boneh, D.: On the effectiveness of address-space randomization. In: Computer and Communications Security, pp. 298–307 (2004)
47. Sovarel, N., Evans, D., Paul, N.: Where's the feeb? the effectiveness of instruction set randomization. In: USENIX Security Conference (2005)
48. Sridhar, S., Shapiro, J.S., Bungale, P.P.: Hdtrans: a low-overhead dynamic translator. SIGARCH Comput. Archit. News 35(1), 135–140 (2007)
49. Sutherland, J.: Business objects in corporate information systems. ACM Comput. Surv. 27(2), 274–276 (1995)
50. Symantec: Internet security threat report. In: http://eval.symantec.com/mktginfo/enterprise/white_papers/ent-whitepaper_symantec_internet_security_threat_report_x_09_2006.en-us.pdf (2006)
51. Thimbleby, H.: Can viruses ever be useful? Computers and Security 10(2), 111–114 (1991)
52. http://info.tikiwiki.org/tiki-index.php (2010)
53. Van Gundy, M., Chen, H.: Noncespaces: Using Randomization to Enforce Information Flow Tracking and Thwart Cross-Site Scripting Attacks. In: Distributed System Security Symposium, pp. 55–67 (2009)
54. Weimer, W., Nguyen, T., Le Goues, C., Forrest, S.: Automatically finding patches using genetic programming. In: International Conference on Software Engineering, pp. 364–367 (2009)
55. Williams, D., Hu, W., Davidson, J.W., Hiser, J.D., Knight, J.C., Nguyen-Tuong, A.: Security through diversity: Leveraging virtual machine technology. IEEE Security and Privacy 7(1), 26–33 (2009)
56. Zeller, A., Hildebrandt, R.: Simplifying and isolating failure-inducing input. IEEE Transactions on Software Engineering 28(2), 183–200 (2002)

Chapter 8
Diversifying the Software Stack Using Randomized NOP Insertion

Todd Jackson, Andrei Homescu, Stephen Crane, Per Larsen,
Stefan Brunthaler, and Michael Franz

Abstract Software monoculture is a significant liability from a computer security perspective. Single attacks can ripple through networks and affect large numbers of vulnerable systems. A simple but unusually powerful idea to solve this problem is to use artificial diversity in software systems. After discussing the design space of introducing artificial diversity, we present an in-depth performance analysis of our own technique: randomly inserting non-alignment NOP instructions. We observe that this technique has a moderate performance impact and demonstrate its real world applicability by diversifying a full system stack.

8.1 Motivation

Networked devices are under constant attack from a wide range of adversaries. Software vulnerabilities such as errors in operating systems, device drivers, shared libraries, and application programs enable most of these attacks. While past research has led to significant results with respect to finding and eliminating vulnerabilities, modern software is exceedingly complex. Consequently, eliminating all possible errors is not commercially feasible.

The existence of residual software errors becomes a significant threat when large numbers of potential targets are simultaneously affected by identical vulnerabilities. Unfortunately, this is the situation today. We currently live in a *software monoculture*: for widely used software, identical binary code runs on millions—or in some cases, sometimes hundreds of millions—of computers. This makes widespread

T. Jackson (✉) • A. Homescu • S. Crane • P. Larsen • S. Brunthaler • M. Franz
Department of Computer Science, University of California, Irvine, CA 92717-3435, USA
e-mail: tmjackso@uci.edu; ahomescu@gmail.com; sjcrane@uci.edu; perl@uci.edu;
stefan@brunthaler.net; franz@uci.edu

S. Jajodia et al. (eds.), *Moving Target Defense II: Application of Game Theory
and Adversarial Modeling*, Advances in Information Security 100,
DOI 10.1007/978-1-4614-5416-8_8, © Springer Science+Business Media New York 2013

exploitation easy and attractive for an attacker, because the same attack vector is likely to succeed on a large number of targets.

A successful software exploit requires two steps. First, an attacker needs to find an entry point through which he or she can inject a malignant payload. Second, the attacker needs to redirect the flow of control on the target computer to the injected payload. In early software attacks, the payload was executable code. For example, in the "classic" buffer overflow attack [3], the stack is overwritten and the return address is changed to point directly to this attack code. This straightforward attack is now prevented on modern platforms through non-executable memory protections such as Microsoft's Data Execution Prevention (DEP), CPU-supported non-executable memory (NX and XD bits), or mandatory code signing such as on Apple's iOS operating system for smartphones and tablets.

Preventing the injection of directly executable code has not stopped all attacks. Instead, attackers have switched to a strategy in which perfectly legitimate code already present on the target computer is used *indirectly*. These attacks "thread through" existing code snippets; names that have been used for attacks of this nature include "arc injection," "borrowed code," [22] and "return-oriented programming" [34].

Unfortunately, these indirect attacks are in practice just as powerful as direct code injection attacks. On virtually every computer that is a potential target, there is sufficient legitimate code already present that the attacker will find any functionality he or she wants to "borrow." Over the past few years, the extent of this threat has slowly revealed itself. It turns out to be extremely difficult to defend against this indirect type of attack using conventional means. This type of attack has been implicated in the Stuxnet worm [26] and the Internet Explorer "Aurora" attack [4].

We present an approach that does not remove software vulnerabilities, but raises the cost to a potential attacker. A central component of our solution is massive-scale binary code diversity, generated automatically by compiler-based techniques. This diversity has the effect that any single attack will succeed only on a small fraction of the installed user base; a large number of different attacks would be necessary to affect many targets and an attacker cannot know which attack variant needs to be directed at which target.

Not only does this remove the problem of viral attacks, in which a single exploit package (such as a worm) ripples through a large number of network-connected hosts in a very short time, but it also makes targeted attacks economically unviable. Our approach also mitigates directed attacks by sophisticated attackers, even those with nation-state scale resources, because even such attackers cannot guess which particular binary version is running on a specific target. Absent an insider who communicates the entirety of the specific binary running on the target to the attacker, the attacker would not know which few of a large number of different attack vectors is going to be successful—and any attempt to brute force this problem by using the many alternative attacks in succession would require substantial time and generate significant amounts of network traffic that is likely to lead to detection.

First, we discuss the design space of introducing artificial software diversity—including relevant related work—and present code-reuse attacks in greater depth (see Sect. 8.2.) Next, we present our rationale for compile-time software diversification, followed by a description of our implementation of randomly inserting non-alignment NOP instructions (see Sect. 8.3.) Finally, we spend most of our efforts on providing a detailed and in-depth analysis of the performance impact of NOP instruction insertion (see Sect. 8.4.) These results indicate that our approach is very practical from a performance perspective and even allows us to diversify the full system stack of a computer, i.e., operating system, system software, and regular applications.

8.2 Background

8.2.1 Software Diversity Design Space

The lifetime of a program generally includes the following stages: development, compilation, linking, deployment, loading and finally running. When designing a diversification technique, two closely related issues must be addressed: how to randomize or transform the software, and at what stages the diversification occurs. Together, these design choices define the software diversification design space and help us understand the strengths and weaknesses of the individual techniques within that space.

The opportunities to introduce diversity at each stage in the software life-cycle are as follows:

During Development. At the outset, diversity can be introduced as the source code level. In 1977, Avizienis and Chen introduced *n-version programming* [2], where multiple independent teams implemented the same program aiming to increase fault tolerance. Linger [23] proposed stochastic source-level transformations to obfuscate software vulnerabilities.

At Compile and Link-Time. Since compilation is an essential and fully automatic step in the software life cycle, it is a natural point to introduce diversification in a way which is mostly transparent to software developers.

Simple transformations introduce diversity in the handling of the call stack to prevent *buffer overflows* [3] or *format string vulnerabilities* [37] from being used by attackers to redirect execution to malicious payloads. Some transformations either place random values—*canaries*—between stack variables and the return address to detect overwriting of the latter [13] or reverse the stack growth direction [33].

The idea of integrity checking return addresses before executing the returns has been extended to include other parts of the program control flow. To ensure *control flow integrity*, the compiler may insert special checks that check if all jump indirect targets belong to a whitelisted set [1] or set a value before an indirect jump transfer and subsequently clear it [8]. Further, the compiler may also try to enforce alignment of all indirect jumps [25,29,40] to prevent code-reuse attacks from using unintended code sequences—a side-effect of variable length instruction encoding.

Finally, the compiler can be used to diversify the instruction mix. Chen et al. modified the compiler to generate code without returns which prevents some (but not all [9, 11]) classes of code-reuse attacks [14]. Jacob et al. [19] introduced the idea of a "superdiversifier," a compiler that performs superoptimization [24] for the purposes of increasing computer security. Our compiler-based approach builds on these ideas by *adding* instructions rather than permuting, substituting or removing the existing instructions. This gives us a significant advantage over prior work: being able to generate a virtually unlimited number of program variants behaving identically towards users but differently towards code-reuse attacks.

An important distinction can be made between techniques that perturb the program implementation and those who rely on integrity checks; the latter adds computation to the program being protected while the former can only affect performance indirectly.

During Deployment. Essential parts of the software stack including operating systems, device drivers, systems software, and applications are available in several different versions. Han et al. [16] studied diversity in off-the-shelf software distributions and found that 98.5% of commonly attacked software have substitutes which are unlikely to be vulnerable to an identical attack.

Additional diversity can be introduced during deployment (and whenever the program is not running) by rewriting program binaries. Bhatkar et al. [7] present an implementation which randomizes the base address of the stack, heap and starting addresses of dynamically linked libraries.

Software updates re-deploy programs partially or in full. With our compiler-based approach to diversification, it is natural to distribute a new program variant or re-diversify during the patching process so attackers cannot learn about a randomized binary by reverse engineering its patches.

At Load-Time. Instead of using a binary rewriter, the base address can be randomized by the operating system loader [30]. Notably, this diversity technique is widely employed by modern operating systems.

Instruction Set Randomization [5, 21, 39] (ISR) is another interesting approach towards introducing diversity at deployment or load time. The idea is to hide the expected instruction encoding from an attacker. Without hardware support for this technique, a virtual machine is necessary to decode the instructions back into a format understood by the physical machine.

At Runtime. Stack protection and general control-flow protection schemes are enforced at this stage. An unexpected control flow transfer usually terminates the program. Owing to the flexibility of a virtual machine, one can re-diversify the program by changing the instruction set encoding as it executes with no additional performance impact [28]. Adaptations of *n*-version programming [6, 12, 35] to runtime systems run multiple versions of the same program in lockstep for the purposes of protecting the programs for attack.

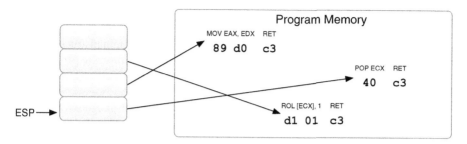

Fig. 8.1 Return-oriented programming attack example, beginning with a POP gadget

It is worth noting that diversification performed after deployment means the same program binaries can be shipped to all users thus remaining compatible with traditional distribution processes. However, if the diversification mechanism is disabled—accidentally or by an attacker—the binary is left unprotected. Pre-deployment approaches, on the other hand, deliver true always on, tamper proof diversification and prevents reverse engineering of patches. However, this approach requires each user to download a uniquely diversified binary thus requiring changes to software distribution and error reporting. Just as local diversification techniques can be compromised, a diversifying software distribution platform can as well. Consequently, a hybrid compile/load-time diversification approach is likely more secure than each approach in isolation.

8.2.2 Return-Oriented Programming and Code-Reuse Attacks

Return-oriented programming (ROP) was initially demonstrated by Shacham [34] in 2007 for the x86 architecture. This technique is a generalization of the return-to-libc attacks first described by Nergal in 2001 [27] and an evolution of the "borrowed code chunks" technique described by Krahmer [22]. Buchanan et al. [10] extended this in 2008 to a generalized version of ROP, which worked on fixed-width instruction set architectures.

In doing so, it evades defenses which protect the stack and other data from being executed. Return-oriented programming requires scanning the target program and/or its libraries for *gadgets*, small executable byte sequences that end in a return instruction. The addresses are then written to the program stack, so that when the current function terminates, it jumps to the first gadget. Because gadgets end in a return instruction, program execution proceeds to the gadget whose address is listed next on the stack. Figure 8.1 shows an example attack in which the POP gadget pointed to by the stack pointer subsequently causes the MOV gadget to be executed, followed by the ROL gadget. Shacham and others have demonstrated that not only is program control flow is possible with careful arrangement of gadgets, but return-to-libc attacks and ROP attacks are both Turing complete [34, 38].

Checkoway et al. [11] demonstrate a return-less approach that thwarts Li et al.'s and Chen et al.'s defenses. Instead of using the conventional approach which depends on RET instructions, Checkoway et al. make use of certain instructions that behave like a return instruction. Similarly, *jump-oriented programming* [9] does not require use of the stack in any way and does not require that gadgets end in the RET instruction.

The original description of ROP used a hand-picked set of gadgets found in a specific version of the C library. Several automated tools [17, 32, 36] scan a given binary, produce a set of useful gadgets, and optionally a payload that uses those gadgets. These tools build a database of gadgets that is, in most cases, Turing-complete. Since these implementations use advanced techniques to combine gadgets (machine learning, SAT solvers), they usually require the entire binary to compute the gadget sets. Hund et al. [17] introduce a language, scanner and compiler for ROP-based exploits which scan the Windows NT kernel and drivers, and builds a database of short gadgets, which form a Turing-complete set. The gadgets are restricted in length to a few bytes, so that the interactions between instructions in a gadget are minimized. The DEPLIB 2.0 ROP payload generator [36] uses a novel approach to generate gadgets. DEPLIB 2.0 expresses the flag, memory and register semantics of instructions in gadgets as boolean expressions, one per output value, and then uses an SMT solver to find gadgets that satisfy logical expressions corresponding to a ROP-based attack. Roemer et al. [31] describe a proof-of-concept exploit language and compiler that generates return-oriented exploits.

8.3 Implementation

In 2010, Franz identified several important paradigm shifts advocating compilation-based introduction of software diversity [15]. Along these lines, we pursued several research directions and published results in a previous installment of a book in this series [20]. In the remainder of this chapter, we present our in-depth evaluation of randomly inserting non-alignment NOP instructions to diversify software. While there are several reasons for inserting NOP instructions per se, we think that the simplicity of this approach is its biggest benefit. A direct consequence of this simplicity is that we maintain program semantics by construction.

8.3.1 System Description

To implement compiler-based diversity we modified LLVM 2.9 and GCC 4.6.2 to introduce the NOP instructions. Our LLVM implementation adds an additional diversifying pass just before final machine code generation. This pass inserts a NOP instruction with some probability before each instruction as it generates the machine instruction stream. Although LLVM with the Clang front end allows

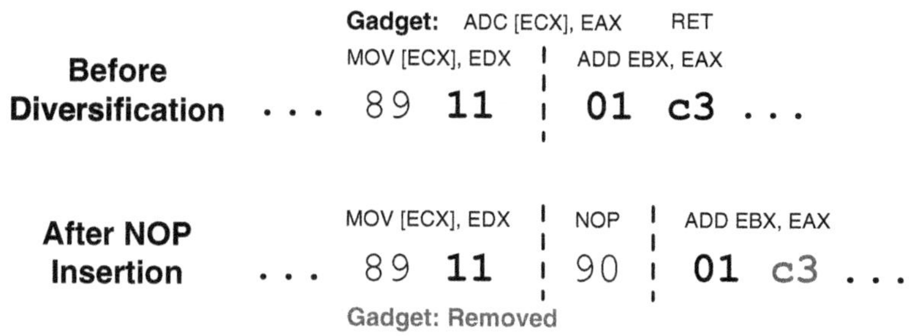

Fig. 8.2 Example of instruction stream changed by inserting a NOP

us to generate diversified binaries for most software, some software relies on GCC-specific idiosyncrasies and is therefore difficult to compile using LLVM. To handle such software, we also add similar modifications to GCC. In the GCC implementation we insert NOPs in GCC's machine independent register transfer language, before instruction selection. This approach is slightly more general, since it allows NOP insertion on architectures other than x86, but the LLVM approach can be extended to support other architectures just as well.

8.3.2 Choosing Candidate NOP Instructions

Many gadgets frequently start in the middle of some proper instruction in the instruction stream. Therefore, most, if not all, of the remaining instructions in the gadget also start somewhere in the middle of an instruction that was properly generated by the compiler. Since x86 instructions have variable lengths and formats which are highly dependent on the first byte of the instruction, i.e., the opcode byte, even a single byte inserted somewhere inside the byte array of a gadget can significantly alter that gadget, if not neutralize it completely (Fig. 8.2).

For instance, inserting one extra byte could cause a following C3 byte, which is the opcode for the return instruction, to be decoded as one of the operands of a previous instruction, eliminating the gadget completely if no other return instruction follows thereafter. Another desirable effect of these instructions is the displacement of all following instructions, moving them forward by an offset that grows as we insert more of these instructions. These *no-operations*, also called NOPs, are instructions of varying lengths that do not modify any processor state.

The x86 instruction set offers many different NOPs of varying lengths, from the 1-byte 0x90 instruction to instructions as long as 16 bytes. To ensure that our approach does not create new gadgets, we have only used 1- and 2-byte instructions where the second byte can be decoded as a NOP operand or as an opcode of an instruction with favorable and well-determined effects. For now, the only values we use for the second byte are opcodes for instructions that make a gadget containing this instruction completely harmless, as seen in Table 8.1.

Table 8.1 Instruction
sequences used as NOPs

Instruction	Encoding	Second byte decoding
NOP	90	–
MOV ESP, ESP	89 E4	IN
MOV EBP, EBP	89 ED	IN
XCHG ESP, ESP	87 E4	IN
XCHG EBP, EBP	87 ED	IN
LEA ESI, [ESI]	8D 36	SS:
LEA EDI, [EDI]	8D 3F	AAS

Initially, our implementation of NOP insertion included the XCHG instructions listed in Table 8.1. Preliminary tests indicated that use of these instructions caused a significant performance overhead. As a result, we removed the XCHG instruction from our implementation and they are not included in the evaluation in Sect. 8.4. The XCHG instructions are grayed out in our table to illustrate this fact.

8.3.3 Randomized Insertion of NOP Instructions

Since we use a randomization-based approach, the NOPs are randomly inserted into the instruction stream of the program at code generation time. The compiler receives a parameter p_{NOP} representing the probability that an instruction will be prepended with a NOP. The parameter p_{NOP} has a direct impact on the diversity and number of available gadgets, but also an inverse impact of the performance of the randomized binary. Therefore, careful selection of this parameter is crucial to obtaining a good balance between performance and diversity.

8.3.4 Applications

Diversification at compile-time allows us to diversify and protect any existing software artifact. One can implement diversification at any level of a system, depending on security requirements and acceptable security trade-offs. For high-security applications, one could diversify the entire operating system, from the kernel up to the application level. If this level of security is too excessive, one might want to only diversify high-risk applications, such as interpreters or web browsers. To demonstrate real-world applications, we compile proof-of-concept applications at each of the following levels:

Operating System Diversity. A diversifying compiler could easily compile any source-based operating system such as Linux or FreeBSD from the ground up. This would protect the entire system stack, and all user-level applications could be diversified on top of this platform, resulting in a secure system at all levels.

System Software Diversity. Diversity at the application level is useful to specifically protect vulnerable applications, such as network facing clients and services. To demonstrate this protection we build diversified versions of Apache and Samba, two frequently attacked network services.

Application Diversity. Since vulnerabilities in systems software, such as interpreters, result in widespread attacks, diversification could help to mitigate this threat. We compile and test diversified versions of Python and the Chromium browser to demonstrate that this is indeed feasible and that the performance impact is moderate. It is worth noting that diversifying a just-in-time compiler, such as Chromium's JavaScript engine, does not result in a significant impact on performance since the code generated at runtime is not diversified.

8.4 Evaluation

8.4.1 Distribution

It used to be the case that software was predominantly shipped on physical media in shrink wrapped boxes. If that was the situation today, the case for compiler-based diversity would have been less strong. Fortunately, this is no longer the case. Since smartphones and tablets often lack media readers, online software distribution platforms (App Stores) have been used instead. The App Store concept has been very successful since it makes it easy for customers to discover and purchase software, easy for developers to deliver updates and prevent illegal copying, and easy for the App Store owner to revoke malicious or vulnerable programs—so successful in fact, that the App Store model is being deployed in general purpose operating systems, too.

Delivering diversified binaries would impact an online distribution platform in two ways: First, the sizes of the binaries to be stored and downloaded by end-users changes. Second, the compilation time is impacted by the added diversification step.

Executable Size. To measure how NOP insertion affects file sizes, we created several diversified binaries by compiling a set of popular software packages. Like our performance evaluation, we compiled several versions of the same program with identical randomization parameters to obtain a representative sample. Since programs are typically transmitted in a compressed format but stored without compression, we study file size increases with and without compression. Files were compressed using an open source implementation of the Burrows-Wheeler algorithm (bzip2) which offers good compression ratios and fast decompression.

Figure 8.3 shows increases in compressed file sizes for seven p_{NOP} values between 1 and 100. The compressed file sizes range from 14 KiB to 29659 KiB and are on average 69% smaller than their uncompressed equivalents which range from 38 KiB to 83393 KiB. Even though compression ratios are necessarily dependent

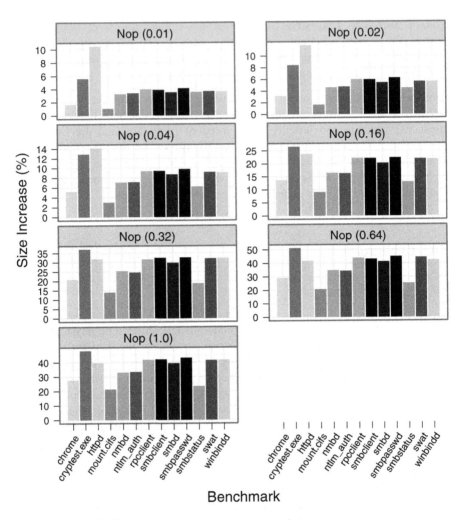

Fig. 8.3 Executable file size increase after `bzip2` compression

on the number of NOP instructions inserted, we observe that the ratios are even more dependent on the particular program being compressed; both smbd and smbpasswd compress to 27% of their original sizes when $p_{NOP} = 0.01$.

NOP insertion caused compressed file sizes to grow between 1% and 51% with mount.cifs compiled with $p_{NOP} = 0.01$ and cryptest compiled with $p_{NOP} = 0.64$ accounting for the smallest and largest increases respectively. Setting p_{NOP} to 0.01 increases compressed sizes between 1 and 10% whereas p_{NOP} at 1 leads to increases between 21% and 48%. Looking at the compressed growth pattern for a single program, we see a curve resembling a logarithmic function with steady increases for small p_{NOP} values which tapers off significantly once p_{NOP} exceeds 0.32. On average, NOP insertion increased file sizes by 20% after compression versus 13% increases before.

For uncompressed binaries, we observe that the effects of NOP insertion are more varied. Diversification with $p_{NOP} = 0.01$ increases file sizes by 0%–38% whereas setting $p_{NOP} = 1$ leads to increases sizes between 21% and 81% with smaller binaries being significantly more sensitive than larger ones. This is to be expected since some binaries, such as chrome, have a large size as a result of containing embedded resources, such as multimedia files, which are unaffected by our diversification.

In summary, compression significantly reduces the extra bandwidth required to distribute diversified binaries to end users.

Build Times. We have conducted timing experiments to measure build times using SPEC CINT2006 and our LLVM-based compiler. NOP insertion increases build time by an average of 0.5% over a normal 187 s build of SPEC CINT2006. We believe that the increase in build time is proportional to the increase in the number of instructions in the binary and that the increase is chiefly due to the larger files that the compiler and linker must write to the disk.

Using LLVM's diagnostic capabilities, we are able to determine that our transformation passes take little time and that LLVM's overall compile time is dominated by other components.

To determine how the build time scaled, we also tested with a development version of the Chromium browser. We found that on a sequential build that normally averages 85 min, a diversified build typically required less than 60 s of extra time. Hence, our testing shows a negligible impact on build time across a broad range of diversification parameters and code sizes.

The results in this section demonstrates that using cloud-based services to build diversified software is indeed practical. Developers and software publishers can use build times for undiversified software as an estimate of how much computing time diversified binaries will require.

In future work, we plan on dramatically reducing build times. Currently, each diversified binary is created by compiling the source code from scratch. Since diversification is the last step in our compilation process, we can cache and reuse the work done at all preceding steps for each additional diversified binary. Because LLVM has a modular design, the required changes are modest. Currently, we emit diversified machine code for each compilation unit. Instead, we could store undiversified LLVM intermediate representation and perform diversification when it is linked and transformed into machine code. In this scheme, only the linking step will need to be repeated to create an unique binary.

8.4.2 Performance

Operating System. To show entire operating system diversity is feasible, we created a completely diversified proof-of-concept Linux installation on a test system with a 2.50 GHz Intel Core 2 Quad Q9300 processor. We used a development

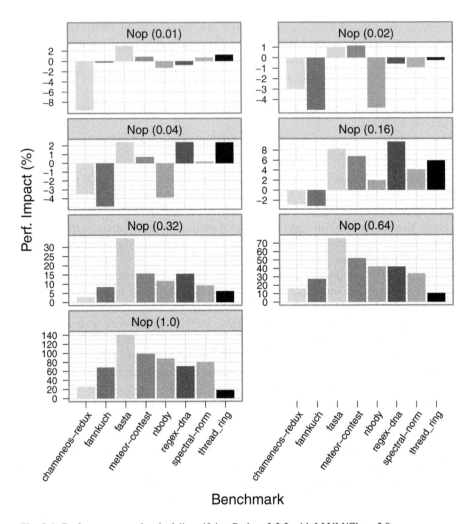

Fig. 8.4 Performance overhead of diversifying Python 3.2.2 with LLVM/Clang 2.9

version of the Linux From Scratch project, which builds all software from source and bootstraps a compiler. Using an automated build tool, we built three diversified Linux From Scratch installations with p_{NOP} set to 0.64. To evaluate the performance of this system, we benchmarked a diversified Apache 2.2.21 with $p_{NOP} = 0.64$ using ApacheBench on the loopback network interface. Averaged over the three diversified builds, we observed a slowdown of 10.5% when serving a 36.1 KiB HTML file and 17.8% when serving a 15.4 KiB GIF image compared to an undiversified operating system (Fig. 8.4).

Table 8.2 Performance slowdown experienced on the Kraken benchmark suite

p_{NOP}	Benchmark category				
	ai (%)	audio (%)	imaging (%)	json (%)	stanford (%)
0.01	−0.2	−7.7	−4.4	3.1	−0.3
0.02	−0.2	−10.8	−5.2	2.5	−0.2
0.04	−0.2	−9.5	−0.8	0.1	0.2
0.16	−0.1	−1.3	−1.8	3.1	1.0
0.32	0.7	−7.7	0.1	8.0	2.2
0.64	0.0	−6.4	−0.4	13.1	5.2
1.00	0.3	−6.2	10.9	28.1	9.4

Python. To illustrate the performance-security trade off, we used our diversifying compilers to build and run several large and well-known software packages. For all tests, we build each package three times with a particular diversification setting and average the results.

We evaluate the performance impact of our transformations using a subset of the C and C++ benchmarks in SPEC CPU2006, Python 3.2.2, Apache 2.2.21 and Chromium 16.0.912.63. These packages illustrate the performance impact of diversification at various layers of the software stack. For compatibility reasons, we build SPEC CPU2006 with both LLVM 2.9 and GCC 4.6.2, Python and Apache with LLVM 2.9 only, and Chromium with GCC 4.6.2. For SPEC CPU2006, we ran our LLVM 2.9 builds on a dual-socket 2.66 GHz Intel Xeon 5150 system, running Linux kernel 3.0.0-13, and our GCC 4.6.2 tests on a dual-socket 2.33 GHz Intel Xeon 5140 system, running Linux kernel 2.6.32-34. Remaining performance tests were run on a dual-socket 2.80 GHz Intel Xeon E5462 system running Linux kernel 3.0.0-13. All performance tests are normalized to a baseline, undiversified build on the same system.

In our performance tests, each executable is run three times. We report the average from the results of each run. For each transformation, we build multiple randomized executables and then run each executable three times to create a better representation of the effect of the transformation. Finally, we average the results of each transformation.

Our Python test suite is composed of the subset of the Computer Language Benchmark Game that is able to run in the default installation of Python 3.2.2. We note that when p_{NOP} is set to 0.01, 0.02, or 0.04, we observe performance overheads within the range of experimental error. However, there are some outliers, such as *chameneos-redux, fannkuch* and *nbody*, which experience speedups of 5%.

When $p_{NOP} = 0.16$, the performance overhead becomes measurably more consistent. Again, *chameneos-redux* and *fannkuch* experience minor speedups. Other tests become slower, with *regex-dna*'s performance degrading by 9%.

As the NOP insertion parameter increases, other tests display sensitivity to the extra NOP instructions in the Python interpreter. The *fasta* test becomes significantly slower, mainly due to the large amount of output the test generates.

Chromium. While the Python test suite is notable because the Python interpreter does not include a just-in-time compiler, we also determine the effect of NOP insertion on software that uses this feature. As a result, our test suite includes running the SunSpider and Kraken JavaScript test suites on Chrome 16. Chrome's JavaScript functionality is notable because it combines V8 and Crankshaft, both just-in-time compilers. As a result, this test shows how another kind of user-facing software will behave when diversified. Note that these results are the scores provided by the benchmark suite itself, and does not include browser startup time or the time to download the test suite.

On the Kraken benchmarks (Table 8.2), we notice a significant performance increase in the *audio* suite. This is largely due to a sub-test, *beat-detection*, executing significantly faster, and pulling the average down. On the other hand, *json* is consistently slower with slowdowns near 3% with $p_{NOP} = 0.01$ and 28% with $p_{NOP} = 1$.

The SunSpider tests (Table 8.3) continue the trend of minor speedups turning into performance overheads that the Kraken tests demonstrated. For example, *3d* shows a 7% performance improvement with $p_{NOP} = 0.01$, but a 10% overhead with $p_{NOP} = 1$. The *crypto* benchmark, on the other hand, demonstrates a consistent overhead from 1% to 9%.

Apache. Our Apache tests involve using ApacheBench to repeatedly request the same file from our test system. We flush the disk cache before running each test. This forces the kernel to warm up the cache during the test as opposed to attempting to ensure that the disk cache has been properly initialized. ApacheBench itself runs on another system, connected to the test system via a 100 Mbit Ethernet connection. During each test, ApacheBench requests the file 10,000 times.

The sample files used in this test are a 36.1 KiB HTML file containing the GNU General Public License version 3, and a 15.4 KiB GIF image.

We note that GIF file throughput (number of requests serviced per second) experiences average decreases by 8%–10% as NOPs are inserted into the Apache binary. However, the averages hide the variance in the test results at each p_{NOP} setting. For example, with $p_{NOP} = 0.01$, the throughput decrease ranges from 7.5% to 10.7%. Interestingly, with $p_{NOP} = 0.64$, the decrease is consistently higher than 10%. This is mostly due to the fact that when $p_{NOP} = 1$, there is little actual diversity because there are no options for random placement of NOP instructions.

For the HTML file, we note that there is little change in throughput, in spite of serving a larger file. This could indicate that Apache has saturated the network link and the resulting throughput changes are not noticeable on the client side. This result indicates that there are some cases where the performance trade off will not be noticeable by common users, especially over a network connection.

SPEC CPU2006. When testing the effects of diversification on the performance of SPEC CPU2006, we tested using both of our diversifying compilers and only compare results generated by the same compiler. For compatibility reasons, we have removed some tests that do not compile with GCC.

Table 8.3 Performance slowdown experienced on the SunSpider JavaScript benchmark suite

P_{NOP}	Benchmark category								
	3d (%)	access (%)	bitops (%)	controlflow (%)	crypto (%)	date (%)	math (%)	regexp (%)	string (%)
0.01	−7.1	−4.2	−1.2	−2.5	1.1	−1.3	−1.9	0.0	−0.2
0.02	−4.0	−3.2	−0.5	−4.2	1.1	−1.3	−0.1	−1.4	0.4
0.04	−4.5	−1.9	−2.7	−1.7	0.9	−1.3	−1.4	−0.9	−0.8
0.16	−2.4	−4.2	0.7	−0.8	3.3	−1.1	−0.3	0.9	0.5
0.32	−2.9	−0.9	−1.4	0.8	3.9	2.7	−2.7	−0.3	3.3
0.64	4.1	4.6	0.6	−5.0	3.0	5.5	−0.1	−0.6	10.5
1.00	10.1	3.9	3.0	1.7	8.5	10.2	−0.3	−1.4	15.8

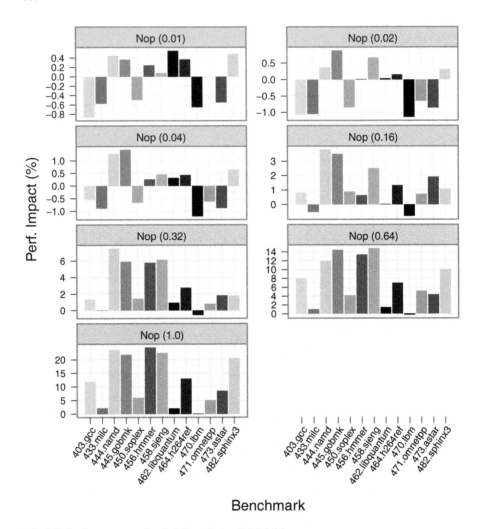

Fig. 8.5 Performance overhead of diversifying GCC 4.6.2

Our diversifying GCC produced no significant difference in the performance of SPEC CPU2006 with $p_{NOP} = 0.01$, 0.02, and 0.04 (Fig. 8.5). Any significant gains or slowdowns at that setting are within experimental error. However, with $p_{NOP} = 0.16$ and up, there is a measurable impact. At that setting, *444.namd* and *445.gobmk* are slower by over 3%, and over 20% with $p_{NOP} = 1$. *470.lbm* is notable in that it is largely unaffected by diversification, showing overheads or speedups less than 1.5% and within experimental error. We also note that there appears to be a bimodal distribution of overheads in this experiment, with one mode centered around tests that are not significantly affected and another centered around tests with measurable, obvious performance overheads.

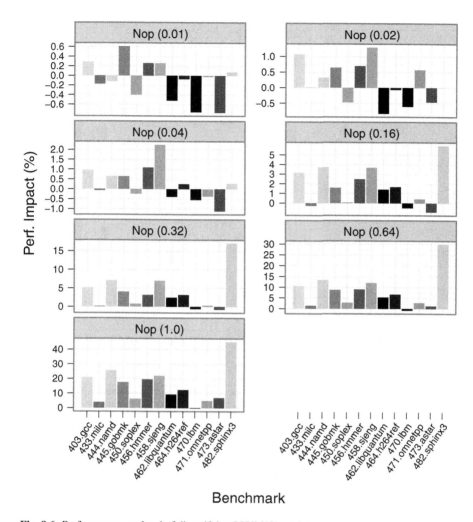

Fig. 8.6 Performance overhead of diversifying LLVM/Clang 2.9

Like the GCC test results, the diversified executables produced no significant difference in overhead when compiled with LLVM 2.9 and diversified with p_{NOP} = 0.01, 0.02, and 0.04 (Fig. 8.6). Also, *470.lbm* remains notable in not being affected by diversification. *482.sphinx3* is significantly slower as p_{NOP} increases past 0.04, and is 45% slower with p_{NOP} = 1. The difference in the overhead in *482.sphinx3* is due to the code analysis and generation methods used in each compiler. In particular, GCC's loop vectorization optimization provides significant performance benefits. Again, in a similarity to the GCC results, we note that the bimodal distribution of performance overhead remains. The primary difference in the distribution is that *462.libquantum* and *464.h264ref* now have 10% overhead with p_{NOP} = 1.

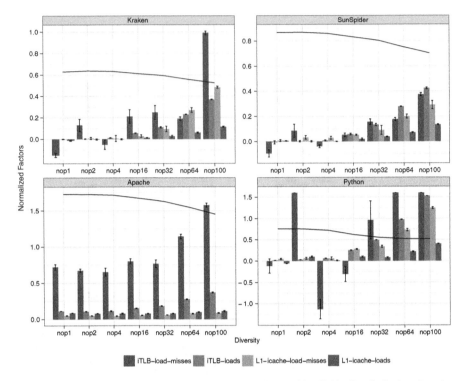

Fig. 8.7 Performance counter measurements on the Kraken and SunSpider JavaScript benchmarks plus a subset of the Computer Language Benchmarks Game suite for Python. For presentation purposes, the height of the bars showing I-TLB load misses for NOP insertion percentages 2, 64 and 100 were reduced. Their values are 6.0, 8.1, and 52.3 respectively

Underlying Causes of Performance Impact. Seeking a better understanding of the slowdowns observed after NOP insertion, we captured a range of execution metrics using hardware performance counters. The collected execution metrics include cycles, branches and instructions executed as well as memory hierarchy events collected separately for instruction and data accesses.

We collected these metrics using the perf tool from the Linux kernel developers. This tool collects the values of various hardware and software counters during the execution of a target process and all its children. The profiling run of each binary was repeated five times to improve measurement accuracy. Like the measurement of program runtimes, we build and measure three randomized versions at each diversification setting and report averages over these.

For brevity, we focus the discussion on hardware metrics which changed substantially in response to NOP insertion. These are loads and misses in the L1 instruction cache and the instruction translation lookaside buffer, I-TLB, which translates virtual addresses to physical ones. We also observe the changes in the cycles-per-instruction ratio, CPI, with increasing amounts of diversity (see Fig. 8.7).

Table 8.4 Throughput slowdown on Apache 2.2.21

p_{NOP}	File transferred	
	sit3-shine.7.gif (%)	gpl-3.0-standalone.html (%)
0.01	9.0	0.0
0.02	9.4	0.0
0.04	9.9	0.0
0.16	11.1	0.0
0.32	9.6	0.0
0.64	11.3	0.0
1.00	10.2	2.5

Looking at the Kraken and SunSpider benchmarks on Chromium, we see that with low NOP insertion probabilities (0.01–0.04) the I-TLB misses fluctuate between 13% and −15% whereas the number of accesses are largely unchanged. On the other hand, higher insertion probabilities (0.16–1.00) all increase the I-TLB accesses between 6% and 43%. Meanwhile, the number of misses increased between 5% and 99%. The pattern repeats itself for instruction cache utilization—accesses and misses stay within the margin of experimental errors for low insertion probabilities whereas probabilities from 0.16 and up increase L1 I-cache accesses and misses by 1%–13% and 3%–48%, respectively. Examining the changes in instructions retired and processor cycles is also instructive. The CPI decreases steadily from 0.63 to 0.53 (Kraken) and from 0.87 to 0.71 (SunSpider), indicating that on average, the added instructions pass through the CPU pipeline in fewer cycles than the rest of the instructions. Since processor designers do not expect compilers to insert NOPs except for alignment purposes, and such instructions are infrequently executed, Intel does not optimize its instruction decoders to recognize and discard any type of NOP instruction [18]. However, the 1-byte NOP instruction (0x90 or XCHG EAX, EAX) does have special hardware support which discards its dependence on the EAX register, thus making it the cheapest NOP instruction.

For Apache, I-TLB misses rise sharply with between 72% and 157% whereas I-TLB loads only increase by 11%–37%. However, the pattern of the increases are not reflected in the throughput slowdowns of Table 8.4. Further, we only observe a moderate increase in I-cache loads (8%–11%) and misses (4%–8%). The number of instructions retired immediately increase to 11% at insertion probability 1% and increase to 38% at insertion probability 100%. Increases in cycle counts range from 8% to 14% meaning that CPI decreases from 1.73 to 1.45.

As previously identified, the Python interpreter shows the highest sensitivity to diversification. Like Apache and Chromium, the CPI decreases steadily from 0.76 to 0.52. However, instruction cache misses increase by 5%–125% and I-TLB misses fluctuate from a factor of −1.14 with $p_{NOP} = 0.04$ to a factor of more than 52 when $p_{NOP} = 1$.

Since Python benchmark runtimes increase by no more than 140%, we believe I-TLB misses are not necessarily the main bottleneck. For instance, it is possible that latency increases from the rise in translation misses (which can be served from caches at lower levels rather than main memory) are hidden or overshadowed by the slowdowns caused by the increase in the L1 instruction cache misses. Further, L2 cache misses across all experiments tend to decrease (except SunSpider which sees a small increase) with increasing insertion probabilities, indicating that the slowdowns are likely caused by the decreased L1 instruction cache utilization. To confirm this theory, our future work will expand the performance studies to include processor specific events relating to instruction fetch, decoding, and pipeline stalls.

8.5 Conclusion

In this chapter we demonstrate the use of NOP insertion for introducing artificial software diversity at compile time. Furthermore, we present results of a careful and detailed analysis of NOP insertion, particularly focusing on its performance impacts and the real-world applicability of our technique. Our data indicates that the performance impact of inserting different kinds of NOP instructions is moderate, and our experiments clearly show that our technique is well suited for use in real-world applications. In addition, our experience with full system stack diversification is promising. Furthermore, our data includes the effects of NOP insertion on build-time and resulting file sizes. The data is important to quantify the effects of moving distribution and diversification to the cloud, as we can estimate the computation, storage, and network data transfer requirements and costs of diversified binaries with necessary precision.

These results are very promising and we expect that subsequent research in this area is going to conclusively show the benefits of NOP insertion from a security perspective. First, we want to quantify the security aspects of our techniques to thwart code-reuse attacks by showing how many pieces of original code "survive" the diversification process in a form usable by an attacker. While we cannot in general know exactly which pieces of code an attacker will use in a code-reuse attack, we envision that a suitable code analysis can compute a conservatively correct estimate of surviving code. Furthermore, such an estimate would also allow us to analyze the performance-security trade off of inserting NOP instructions to guide not only the selection of diversification settings, but also the development of additional transformation techniques.

Acknowledgements Parts of this effort have been sponsored by the Defense Advanced Research Projects Agency (DARPA) under agreement number D11PC20024, and by a generous gift by Google.

The U.S. Government is authorized to reproduce and distribute reprints for Governmental purposes notwithstanding any copyright annotation thereon. Any opinions, findings, and conclusions or recommendations expressed here are those of the authors and do not necessarily reflect the views of DARPA or Google.

References

1. M. Abadi, M. Budiu, Ú. Erlingsson, and J. Ligatti. Control-flow integrity principles, implementations, and applications. *ACM Transactions on Information System Security*, 13:4:1–4:40, 2009.
2. A. Avizienis and L. Chen. On the implementation of n-version programming for software fault tolerance during execution. In *Proceedings of the International Computer Software and Applications Conference*, pages 149–155, 1977.
3. Aleph One. Smashing the stack for fun and profit. *Phrack Magazine*, Issue 49, 1996.
4. Internet Explorer "Aurora" Attack, 2010. (CVE-2010-0249).
5. E.G. Barrantes, D.H. Ackley, S. Forrest, and D. Stefanović. Randomized Instruction Set Emulation. *ACM Transactions on Information and System Security*, 8(1):3–40, 2005.
6. D. Bruschi, L. Cavallaro, and A. Lanzi. Diversified process replicae for defeating memory error exploits. In *Proceedings of the International Workshop on Information Assurance*, pages 434–441, 2007.
7. S. Bhatkar, D.C. DuVarney, and R. Sekar. Address Obfuscation: An Efficient Approach to Combat a Broad Range of Memory Error Exploits. In *Proceedings of the 12th USENIX Security Symposium*, pages 105–120, 2003.
8. T. Bletsch, X. Jiang, and V. Freeh. Mitigating code-reuse attacks with control-flow locking. In *Proceedings of the 27th Annual Computer Security Applications Conference*, pages 353–362. ACM, 2011.
9. T. Bletsch, X. Jiang, V. Freeh, and Z. Liang. Jump-oriented programming: a new class of code-reuse attack. In *Proceedings of the 6th ACM Symposium on Information, Computer and Communications Security*, pages 30–40, 2011.
10. E. Buchanan, R. Roemer, H. Shacham, and S. Savage. When good instructions go bad: generalizing return-oriented programming to RISC. In *Proceedings of the 15th ACM Conference on Computer and Communications Security*, pages 27–38, 2008.
11. S. Checkoway, L. Davi, A. Dmitrienko, A. Sadeghi, H. Shacham, and M. Winandy. Return-Oriented Programming without Returns. In *Proceedings of the 17th ACM Conference on Computer and Communications Security*, pages 559–72, 2010.
12. B. Cox, D. Evans, A. Filipi, J. Rowanhill, W. Hu, J. Davidson, J. Knight, A. Nguyen-Tuong, and J. Hiser. N-variant systems: A Secretless Framework for Security through Diversity. In *Proceedings of the 15th USENIX Security Symposium*, pages 105–120, 2006.
13. C. Cowan, C. Pu, D. Maier, J. Walpole, P. Bakke, D. Beattie, A. Grier, P. Wagle, Q. Zhang, and H. Hinton. StackGuard: Automatic Adaptive Detection and Prevention of Buffer-Overflow Attacks. In *Proceedings of the 7th USENIX Security Symposium*, pages 63–78, 1998.
14. P. Chen, X. Xing, H. Han, B. Mao, and L. Xie. Efficient Detection of the Return-oriented Programming Malicious Code. In *Proceedings of the 6th International Conference on Information Systems Security*, pages 140–155, 2010.
15. M. Franz. E unibus pluram: Massive-Scale Software Diversity as a Defense Mechanism. In *Proceedings of the 2010 Workshop on New Security Paradigms*, NSPW '10, pages 7–16, New York, NY, USA, 2010. ACM.
16. Jin Han, Debin Gao, and Robert H. Deng. On the effectiveness of software diversity: A systematic study on real-world vulnerabilities. In *Proceedings of the 6th International Conference on Detection of Intrusions and Malware, and Vulnerability Assessment*, pages 127–146, 2009.
17. R. Hund, T. Holz, and F.C. Freiling. Return-oriented rootkits: Bypassing kernel code integrity protection mechanisms. In *Proceedings of the 18th USENIX Security Symposium*, pages 383–398, 2009.
18. Intel Corporation. Intel 64 and IA-32 architectures optimization reference manual.
19. M. Jacob, M. Jakubowski, P. Naldurg, C. Saw, and R. Venkatesan. The superdiversifier: Peephole individualization for software protection. In K. Matsuura and E. Fujisaki, editors, *Advances in Information and Computer Security*, volume 5312 of *Lecture Notes in Computer Science*, pages 100–120. Springer Berlin / Heidelberg, 2008.

20. Todd Jackson, Babak Salamat, Andrei Homescu, Karthikeyan Manivannan, Gregor Wagner, Andreas Gal, Stefan Brunthaler, Christian Wimmer, and Michael Franz. Compiler-generated software diversity. In Sushil Jajodia, Anup K. Ghosh, Vipin Swarup, Cliff Wang, and X. Sean Wang, editors, *Moving Target Defense*, volume 54 of *Advances in Information Security*, pages 77–98. Springer New York, 2011.
21. G.S. Kc, A.D. Keromytis, and V. Prevelakis. Countering Code-Injection Attacks with Instruction-Set Randomization. In *Proceedings of the 10th ACM Conference on Computer and Communications Security*, pages 272–280, 2003.
22. S. Krahmer. x86-64 buffer overflow exploits and the borrowed code chunks exploitation techniques. 2005. http://www.suse.de/~krahmer/no-nx.pdf.
23. Richard C. Linger. Systematic generation of stochastic diversity as an intrusion barrier in survivable systems software. In *Proceedings of the Thirty-Second Annual Hawaii International Conference on System Sciences*, pages 3062–, 1999.
24. H. Massalin. Superoptimizer: a look at the smallest program. In *Proceedings of the Second International Conference on Architectual Support for Programming Languages and Operating Systems*, pages 122–126, 1987.
25. S. McCamant and G. Morrisett. Evaluating SFI for a CISC architecture. In *Proceedings of the 15th USENIX Security Symposium*, pages 209–224, 2006.
26. A. Matrosov, E. Rodionov, D. Harley, and J. Malcho. Stuxnet Under the Microscope, 2010. http://go.eset.com/us/resources/white-papers/Stuxnet_Under_the_Microscope.pdf. Accessed 01/09/2012.
27. Nergal. The advanced return-into-lib(c) exploits: PaX case study. *Phrack Magazine*, Issue 58, 2001.
28. Anh Nguyen-Tuong, Andrew Wang, Jason D. Hiser, John C. Knight, and Jack W. Davidson. On the effectiveness of the metamorphic shield. In *Proceedings of the Fourth European Conference on Software Architecture: Companion Volume*, pages 170–174, 2010.
29. K. Onarlioglu, L. Bilge, A. Lanzi, D. Balzarotti, and E. Kirda. G-free: defeating return-oriented programming through gadget-less binaries. In *Proceedings of the 26th Annual Computer Security Applications Conference*, pages 49–58, 2010.
30. PaX. *Homepage of The PaX Team*, 2009. http://pax.grsecurity.net.
31. R. Roemer, E. Buchanan, H. Shacham, and S. Savage. Return-oriented programming: Systems, languages, and applications. *ACM Transactions in Information and Systems Security*, 2011. To appear.
32. E. J. Schwartz, T. Avgerinos, and D. Brumley. Q: Exploit Hardening Made Easy. In *Proceedings of the 20th USENIX Security Symposium*, 2011.
33. B. Salamat, A. Gal, and M. Franz. Reverse Stack Execution in a Multi-Variant Execution Environment. In *Workshop on Compiler and Architectural Techniques for Application Reliability and Security*, 2008.
34. H. Shacham. The Geometry of Innocent Flesh on the Bone: Return-into-libc without Function Calls (on the x86). In *Proceedings of the 14th ACM Conference on Computer and Communications Security*, pages 552–561, 2007.
35. B. Salamat, T. Jackson, G. Wagner, C. Wimmer, and M. Franz. Run-Time Defense against Code Injection Attacks using Replicated Execution. *IEEE Transactions on Dependable and Secure Computing*, 2011.
36. P. Sole. Hanging on a ROPe. In *ekoParty Security Conference*, 2010. http://www.immunitysec.com/downloads/DEPLIB20_ekoparty.pdf.
37. scut / team teso. Exploiting Format String Vulnerabilities. 2001. http://crypto.stanford.edu/cs155/papers/formatstring-1.2.pdf.
38. M. Tran, M. Etheridge, T. Bletsch, X. Jiang, V. W. Freeh, and P. Ning. On the Expressiveness of Return-into-libc Attacks. In *Proceedings of the 14th Interntional Symposium on Recent Advances in Intrusion Detection*, 2011.

39. D. W. Williams, W. Hu, J. W. Davidson, J. Hiser, J. C. Knight, and A. Nguyen-Tuong. Security through diversity: Leveraging virtual machine technology. *IEEE Security & Privacy*, 7(1): 26–33, 2009.
40. B. Yee, D. Sehr, G. Dardyk, J. B. Chen, R. Muth, T. Ormandy, S. Okasaka, N. Narula, and N. Fullagar. Native client: A sandbox for portable, untrusted x86 native code. In *IEEE Symposium on Security and Privacy*, pages 79–93, 2009.

Chapter 9
Practical Software Diversification Using In-Place Code Randomization

Vasilis Pappas, Michalis Polychronakis, and Angelos D. Keromytis

Abstract The wide adoption of non-executable page protections has given rise to attacks that employ return-oriented programming (ROP) to achieve arbitrary code execution without the injection of any code. Existing defenses against ROP exploits either require source code or symbolic debugging information, or impose a significant runtime overhead, which limits their applicability for the protection of third-party applications. Aiming for a practical mitication against ROP attacks, we introduce *in-place code randomization*, a software diversification technique that can be applied directly on third-party software. Our method uses various narrow-scope code transformations that can be applied statically, without changing the location of basic blocks, allowing the safe randomization of stripped binaries even with partial disassembly coverage. We demonstrate how in-place code randomization can prevent the exploitation of vulnerable Windows 7 applications, including Adobe Reader, as well as the automated construction of reliable ROP payloads.

9.1 Introduction

Attack prevention technologies based on the No eXecute (NX) memory page protection bit, which prevent the execution of malicious code that has been injected into a process, are now supported by most recent CPUs and operating systems [49]. The wide adoption of these protection mechanisms has given rise to a new exploitation technique, widely known as *return-oriented programming* (ROP) [61], which allows an attacker to circumvent non-executable page protections without injecting any code. Using return-oriented programming, the attacker can link together small fragments of code, known as *gadgets*, that already exist in the

V. Pappas (✉) • M. Polychronakis • A.D. Keromytis
Department of Computer Science, Columbia University, New York,
NY 10027-7003, USA
e-mail: vpappas@cs.columbia.edu; mikepo@cs.columbia.edu; angelos@cs.columbia.edu

S. Jajodia et al. (eds.), *Moving Target Defense II: Application of Game Theory and Adversarial Modeling*, Advances in Information Security 100, DOI 10.1007/978-1-4614-5416-8_9, © Springer Science+Business Media New York 2013

process image of the vulnerable application. Each gadget ends with an indirect control transfer instruction, which transfers control to the next gadget according to a sequence of gadget addresses injected on the stack or some other memory area. In essence, instead of injecting binary code, the attacker injects just data, which include the addresses of the gadgets to be executed, along with any required data arguments.

Several research works have demonstrated the great potential of return-oriented programming for bypassing defenses such as read-only memory [21], kernel code integrity protections [40], and non-executable memory implementations in mobile devices [29] and operating systems [67, 70–72]. Consequently, it was only a matter of time for ROP to be employed in real-world attacks. Recent exploits against popular applications, such as the ubiquitous Adobe Reader for Windows, use ROP code to bypass exploit mitigations that are enabled even in the latest OS versions, including Windows 7 SP1. ROP exploits are included in the most common exploit packs, and are actively used in the wild for mounting drive-by download attacks [14, 56].

9.1.1 Existing Defenses

Attackers are able to a priori pick the right code pieces before launching a ROP attack because parts of the code image of the vulnerable application remain static across different installations. Address space layout randomization (ASLR) [49] is meant to prevent this kind of code reuse by randomizing the locations of the executable segments of a running process. However, in both Linux and Windows, parts of the address space do not change due to executables with fixed load addresses [34], or shared libraries incompatible with ASLR [72]. Furthermore, in some exploits, the base address of a DLL can be either calculated dynamically through a leaked pointer [46, 60, 70], or brute-forced [62].

Other defenses against code-reuse attacks complementary to ASLR include compiler extensions [47, 54], code randomization [16, 33, 43], control-flow integrity [11], and runtime solutions [22, 26, 27]. In practice, though, most of these approaches are almost never applied for the protection of the COTS software currently targeted by ROP attacks, either due to the lack of source code or debugging information, or due to their increased overhead. In particular, from the above techniques, those that operate directly on compiled binaries, e.g., by permuting the order of functions [16, 43] or through binary instrumentation [11], require precise and complete extraction of all code and data in the executable sections of the binary. This is possible only if the corresponding symbolic debugging information is available, which however is typically stripped from production binaries.

On the other hand, techniques that do work on stripped binary executables using dynamic binary instrumentation [22, 26, 27], incur a significant runtime overhead that limits their adoption. These defenses are based on monitoring either the frequency of ret instructions [22, 26], or the integrity of the stack [27]. At the same time, instruction set randomization (ISR) [13,42] cannot prevent code-reuse attacks, and current implementations also rely on heavyweight runtime instrumentation or code emulation frameworks.

9.1.2 In-Place Code Randomization

Starting with the goal of a practical mitigation against the recent spate of ROP attacks, in this paper we present a novel code randomization method that can harden third-party applications against return-oriented programming. Our approach is based on narrow-scope modifications in the code segments of executables using an array of code transformation techniques, to which we collectively refer as *in-place code randomization* [55]. These transformations are applied statically, in a conservative manner, and modify only the code that can be safely extracted from compiled binaries, without relying on symbolic debugging information. By preserving the length of instructions and basic blocks, these modifications do not break the semantics of the code, and enable the randomization of stripped binaries even without complete disassembly coverage.

The goal of this diversification process is to eliminate or probabilistically modify as many of the gadgets that are available in the address space of a vulnerable process as possible. Since ROP code relies on the correct execution of all chained gadgets, altering the outcome of even a few of them will likely render the ROP code ineffective. The introduced uncertainty raises the bar for the construction of reliable ROP code, as attackers cannot safely assume that a given gadget will perform the intended computation. By randomly choosing and applying different transformations in each instance of the protected application, an attacker will not always be able to choose a safe subset of gadgets that will always remain unchanged.

Still, although quite effective as a standalone mitigation, in-place code randomization is not meant to be a complete prevention solution against ROP exploits, as it offers probabilistic protection and thus cannot deliver any protection guarantees. However, it can be applied in tandem with existing randomization techniques to increase process diversification. This is facilitated by the practically zero overhead of the applied transformations, and the ease with which they can be applied on existing third-party executables.

In the rest of this chapter, we provide some background information on return-oriented programming, discuss the principles of in-place code randomization and the code transformations on which it is based, and present some experimental results using publicly available ROP exploits and automated ROP code generation toolkits.

9.2   From Return-to-Libc to Return-Oriented Programming

9.2.1   Code-Reuse Attacks

The introduction of non-executable memory page protections in popular OSes, even for CPUs that do not support the No eXecute (NX) bit, led to the development of the return-to-libc exploitation technique [28]. Using this method, a memory corruption vulnerability can be exploited by transferring control to code that already exists in the address space of the vulnerable process. By jumping to the beginning of a library function such as `system()`, the attacker can for example spawn a shell without the need to inject any code.

Frequently though, especially for remote exploitation, calling a single function is not enough. In these cases, multiple return-to-libc calls can be "chained" together by ensuring that before returning from one function to the next one, the stack pointer has been correctly adjusted to the beginning of the prepared stack frame for the next call. For instance, for a function with two arguments, this can be achieved by first returning to a short instruction sequence such as `pop reg; pop reg; ret;` found anywhere within the executable part of the process image [52,53]. The `pop` instructions adjust the stack pointer beyond the arguments of the previously executed function (one `pop` for each argument), and then `ret` transfers control to the next chained function. This approach, however, is not applicable in cases where the function arguments need to be passed through registers. In that case, a few short instruction sequences ending with a `ret` instruction can be chained directly to set the proper registers with the desired arguments, before calling the library function [44].

9.2.2   Return-Oriented Programming

In the above code-reuse techniques, the executed code consists of one or a few short instruction sequences followed by a large block of code belonging to a library function. Hovav Shacham demonstrated that using only a carefully selected set of short instruction sequences ending with a `ret` instruction, known as *gadgets*, it is possible to achieve arbitrary computation, obviating the need for calling library functions [61]. This powerful technique, dubbed *return-oriented programming*, in essence gives the attacker the same level of flexibility offered by arbitrary code injection without injecting any code at all—the injected payload comprises just a sequence of gadget addresses intermixed with any necessary data arguments.

In a typical ROP exploit, the attacker needs to control both the program counter and the stack pointer: the former for executing the first gadget, and the latter for allowing its `ret` instruction to transfer control to subsequent gadgets. Depending on the vulnerability, if the ROP payload is injected in a memory area other than the stack, e.g., the heap, then the stack pointer must first be adjusted to the beginning of the payload through a stack pivot [31, 72]. In a follow up work [20], Checkoway et al. demonstrated that the gadgets used in a ROP exploit need not necessarily end with a ret instruction, but with any other indirect control transfer instruction. This also allows the use of any general purpose register in place of the stack pointer as an "index" register for controlling the execution of the gadgets, bypassing any protections based on stack integrity.

Almost a decade after the introduction of the return-to-libc technique [28], the wide adoption of NX-based exploit mitigations in popular OSes sparked a new interest in more advanced forms of code-reuse attacks. The introduction of return-oriented programming [61] and its advancements [17, 19–21, 29, 40, 59, 66, 71, 72] led to its adoption in real-world attacks [14, 56]. ROP exploits are facilitated by the lack of complete address space layout randomization in both Linux [34], and Windows [41, 72], which otherwise would prevent or at least hinder [62] these attacks.

The ROP code implementations used in recent exploits against Windows applications is mostly based on gadgets ending with ret instructions, which conveniently manipulate both the program counter and the stack pointer, although a couple of gadgets ending with call or jmp are also used for calling library functions. In all publicly available Windows exploits so far, attackers do not have to rely on a fully ROP-based implementation for the whole malicious code that needs to be executed after triggering a memory corruption vulnerability. Instead, ROP code is used only as a first stage for bypassing DEP [49]. Typically, once control flow has been hijacked, the ROP code allocates a memory area with write and execute permissions by calling a library function like VirtualAlloc, copies into it some plain shellcode included in the attack vector, and finally jumps to the copied shellcode which now has execute permission [31].

9.3 Approach

In-place code randomization is based on the randomization of the code sections of binary executable files (both libraries and executables) that are part of third-party applications, using an array of binary code transformation techniques. The objective of this randomization process is to break the code semantics of the gadgets that are present in the executable memory segments of a running process, without affecting the semantics of the actual program code.

The execution of a gadget has a certain set of consequences to the CPU and memory state of the exploited process. The attacker chooses how to link the different gadgets together based on which registers, flags, or memory locations each gadget modifies, and in what way. Consequently, the execution of a subsequent gadget depends on the outcome of all previously executed gadgets. Even if the execution of a single gadget has a different outcome than the one anticipated by the attacker, then this will affect the execution of all subsequent gadgets, and it is likely that the logic of the malicious return-oriented code will be severely impacted.

9.3.1 Why In-Place?

The concept of software diversification [23] is the basis for a wide range of protections against the exploitation of memory corruption vulnerabilities. Besides address space layout randomization [49], many techniques focus on the internal randomization of the code segments of executable, and can be combined with ASLR to increase process diversity [33]. Metamorphic transformations [68]. Such as the interspersion of ineffectual instructions throughout the code, can shift gadgets from their original offsets and alter many of their instructions, rendering them unusable. Another simpler and probably more effective approach is to rearrange existing blocks of code either at the function level [5, 15, 16, 43], or with finer granularity, at the basic block level [6, 7]. If all blocks of code are reordered so that no one resides at its original location, then all the offsets of the gadgets that the attacker would assume to be present in the code sections of the process will now correspond to completely different code.

These transformations require a precise view of all the code and data objects contained in the executable sections of a PE file, including their cross-references, as existing code needs to be shifted or moved. Due to computed jumps and intermixed data [45], complete disassembly coverage is possible only if the binary contains relocation and symbolic debugging information (e.g., PDB files) [43, 58, 65]. Unfortunately, debugging information is typically stripped from release builds for compactness and intellectual property protection.

For Windows software, in particular, PE files (both DLL and EXE) usually do retain relocation information even if no debugging information has been retained [63]. The loader needs this information in case a DLL must be loaded at an address other than its preferred base address, e.g., because another library has already been mapped to that location. or for ASLR. In contrast to Linux shared libraries and PIC executables, which contain position-independent code and can be easily loaded in a arbitrary location within a process' address space, Windows binaries contain absolute addresses, e.g., as immediate instruction operands or initialized data pointers, that are valid only if the executable has been loaded at its preferred base address. The `.reloc` section of PE files contains a list of offsets relatively to each PE section that correspond to all absolute addresses at which a delta value needs to be added in case the actual load address is different [57].

Relocation information *alone*, however, does not suffice for extracting a complete view of the code within the executable sections of a PE file [7, 65]. Without the symbolic debugging information contained in PDB files, although the location of objects that are reached *only* via indirect jumps *can* be extracted from relocation information, their actual type—code or data—still remains unknown. In some cases, the actual type of these objects could be inferred using heuristics based on constant propagation, but such methods are usually prone to misidentifications of data as code and vice versa. Even a slight shift or size increase of a single object within a PE section will incur cascading shifts to its following objects. Typically, an unidentified object that actually contains code will include PC-relative branches to other code objects. In the absence of the debugging information contained in PDB files, moving such an unidentified code block (or any of its relatively referenced objects) without fixing the immediate displacement operands of all its relative branch instructions that reference other objects, will result to incorrect code.

Given the above constraints, we choose to use only binary code transformations that do not alter the size and location of code and data objects within the executable, allowing the randomization of third-party PE files *without* symbolic debugging information. Although this restriction does not allow us to apply extensive code transformations like basic block reordering or metamorphism, we can still achieve partial code randomization using narrow-scope modifications that can be *safely* applied even without complete disassembly coverage. This can be achieved through slight, in-place code modifications to the correctly identified parts of the code, that do not change the overall structure of basic blocks or functions, but which are enough to alter the outcome of short instruction sequences that can be used as gadgets.

9.3.2 Code Extraction and Modification

Although completely accurate disassembly of stripped x86 binaries is not possible, state-of-the-art disassemblers achieve decent coverage for code generated by the most commonly used compilers, using a combination of different disassembly algorithms [45], the identification of specific code constructs [35], and simple data flow analysis [36]. For our prototype implementation, we use IDA Pro [38] to extract the code and identify the functions of PE executables. IDA Pro is effective in the identification of function boundaries, even for functions with non-contiguous code and extensive use of basic block sharing [39], and also takes advantage of the relocation information present in Windows DLLs.

Typically, however, without the symbolic information of PDB files, a fraction of the functions in a PE executable are not identified, and parts of code remain undiscovered. Our code transformations are applied conservatively, only on parts of the code for which we can be confident that have been accurately disassembled. For instance, IDA Pro speculatively disassembles code blocks that are reached only through computed jumps, taking advantage of the relocation information contained in PE files. However, we do not enable such heuristic code extraction methods in order to avoid any disastrous modifications due to potentially misidentified code. In practice, for the code generated by most compilers, relocation information also ensures that the correctly identified basic blocks have no entry point other than their first instruction. Similarly, some transformations that rely on the proper identification of functions are applied only on the code of correctly recognized functions. Our implementation is separate from the actual code extraction framework used, which means that IDA Pro can be replaced or assisted by alternative code extraction approaches [37, 51, 65], providing better disassembly coverage.

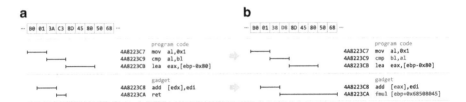

Fig. 9.1 Example of atomic instruction substitution. The equivalent, but different form of the cmp instruction does not change the original program code (**a**), but renders the non-intended gadget unusable (**b**)

After the code extraction phase is complete, disassembled instructions are first converted to our own internal representation, which holds additional information such as any implicitly used registers, and the registers and flags read or written by the instruction. For correctness, we also track the use of general purpose registers even in floating point, MMX, and SSE instructions. Although these type of instructions have their own set of registers, they do use general purpose registers for memory references (e.g., as the fmul instruction in Fig. 9.1). We then proceed and apply the in-place code transformations discussed in the following section. These are applied only on the parts of the executable segments that contain (intended or unintended [61]) instruction sequences that can be used as gadgets. As a result of some of the transformations, instructions may be moved from their original locations within the same basic block. In these cases, for instructions that contain an absolute address in some of their operands, the corresponding entries in the .reloc sections of the randomized PE file are updated with the new offsets where these absolute addresses are now located.

9.3.3 Deployment

Our publicly-available prototype implementation, called orp, can be used to generate randomized instances of existing applications. Orp processes each PE file individually, and generates multiple randomized copies that can then replace the original. To relieve users of the burden of installing and running orp, as part of our future work we also plan to create a web service that will allow the submission of executables for randomization.

Given the complexity of the analysis required for generating a set of randomized instances of an input file (in the order of a few minutes on average for the PEs used in our tests), orp can be used for the off-line generation of a pool of randomized PE files for a given application. Note that for most of the tested Windows applications, only some of the DLLs need to be randomized, as the rest are usually ASLR-enabled (although they can also be randomized for increased protection). In a production deployment, a system service or a modified loader can then pick a different randomized version of the required DLLs and executables each time the application is launched, following the same way of operation as tools like EMET [48].

9.4 In-Place Code Transformations

In this section we present in detail the different code transformations used for in-place code randomization. Although some of the transformations such as instruction reordering and register reassignment are also used by compilers and polymorphic code engines for code optimization [12] and obfuscation [68], applying them at the binary level—without having access to the higher-level structural and semantic information available in these settings—poses significant challenges.

9.4.1 Atomic Instruction Substitution

One of the basic concepts of code obfuscation and metamorphism [68] is that the exact same computation can be achieved using a countless number of different instruction combinations. When applied for code randomization, substituting the instructions of a gadget with a functionally-equivalent—but different—sequence of instructions would not affect any ROP code that uses that gadget, since its outcome would be the same. However, by modifying the instructions of the original program code, this transformation in essence modifies certain bytes in the code image of the program, and consequently, can drastically alter the structure of non-intended instruction sequences that overlap with the substituted instructions.

Many of the gadgets used for return-oriented programming consist of unaligned instructions that have not been emitted by the compiler, but which happen to be present in the code image of the process due to the density and variable-length nature of the x86 instruction set. In the example of Fig. 9.1a, the actual code generated by the compiler consists of the instructions mov; cmp; lea; starting at byte B0.[1] However, when disassembling from the next byte, a useful non-intended gadget ending with ret is found.

Compiled code is highly optimized, and thus the replacement of even a single instruction in the original program code usually requires either a longer instruction, or a combination of more than one instruction, for achieving the same purpose. Given that our aim is to randomize the code of stripped binaries, even a slight

---

[1] The code of all examples throughout this chapter comes from icucnv36.dll, included in Adobe Reader v9.3.4. This DLL was used for the ROP code of a DEP-bypass exploit for CVE-2010-2883 [1] (see Table 9.2).

increase in the size of a basic block is not possible, which makes the most commonly used instruction substitution techniques unsuitable for our purpose.

In certain cases though, it is possible to replace an instruction with a single, functionally-equivalent instruction of the *same* length, thanks to the flexibility offered by the extensive x86 instruction set. Besides obvious candidates based on replacing addition with negative subtraction and inversely, there are also some instructions that come in different forms, with different opcodes, depending on the supported operand types. For example, add r/m32,r32 stores the result of the addition in a register *or* memory operand (r/m32), while add r32,r/m32 stores the result in a register (r32). Although these two forms have different opcodes, the two instructions are equivalent when both operands happen to be registers. Many arithmetic and logical instructions have such dual equivalent forms, while in some cases there can be up to five equivalent instructions (e.g., test r/m8,r8, or r/m8,r8, or r8, r/m8, and r/m8,r8, and r8,r/m8, affect the flags of the EFLAGS register in the same way when both operands are the *same* register). In our prototype implementation we use the sets of equivalent instructions used in Hydan [30], a tool for hiding information in x86 executables, with the addition of one more set that includes the equivalent versions of the xchg instruction.

As shown in Fig. 9.1b, both operands of the cmp instruction are registers, and thus it can be replaced by its equivalent form, which has different opcode and ModR/M bytes [10]. Although the actual program code does not change, the ret instruction that was "included" in the original cmp instruction has now disappeared, rendering the gadget unusable. In this case, the transformation completely *eliminates* the gadget, and thus will be applied in all instances of the randomized binary. In contrast, when a substitution does not affect the gadget's final indirect jump, then it is applied probabilistically.

9.4.2 Instruction Reordering

In certain cases, it is possible to reorder the instructions of small self-contained code fragments without affecting the correct operation of the program. This transformation can significantly impact the structure of non-intended gadgets, but can also break the attacker's assumptions about gadgets that are part of the actual code.

9.4.2.1 Intra Basic Block Reordering

The actual instruction scheduling chosen during the code generation phase of a compiler depends on many factors, including the cost of instructions in cycles, and the applied code optimization techniques [12]. Consequently, the code of a basic block is often just one among several possible instruction orderings that are all equivalent in terms of correctness. Based on this observation, we can partially modify the code within a basic block by reordering some of its instructions according to an alternative instruction scheduling.

The basis for deriving an alternative instruction scheduling is to determine the ordering relationships among the instructions, which must always be satisfied to maintain code correctness. The *dependence graph* of a basic block represents the instruction interdependencies that constrain the possible instruction schedules [50]. Since a basic block contains straight-line code, its dependence graph is a directed acyclic graph with machine instructions as vertices, and dependencies between instructions as edges. We apply dependence analysis on the code of disassembled basic blocks to build their dependence graph using an adaptation of a standard dependence DAG construction algorithm [50, Fig. 9.6] for machine code. Applying dependence analysis directly on machine code requires a careful treatment of the dependencies between x86 instructions. Compared to the analysis of code expressed in an intermediate representation form, this includes the identification of data dependencies not only between register and memory operands, but also between CPU flags and implicitly used registers and memory locations.

For each instruction i, we derive the sets $use[i]$ and $def[i]$ with the registers used and defined by the instruction. Besides register operands and registers used as part of effective address computations, this includes any implicitly used registers. For example, the *use* and *def* sets for pop eax are $\{esp\}$ and $\{eax, esp\}$, while for rep stosb[2] are $\{ecx, eax, edi\}$ and $\{ecx, edi\}$, respectively. We initially assume that all instructions in the basic block depend on each other, and then check each pair for read-after-write (RAW), write-after-read (WAR), and write-after-write (WAW) dependencies. For example, i_1 and i_2 have a RAW dependency if any of the following is true: (a) $def[i_1] \cap use[i_2] \neq \emptyset$, (b) the destination operand of i_1 and the source operand of i_2 are both a memory location, (c) i_1 writes at least one flag read by i_2.

Note that condition (b) is quite conservative, given that i_2 will actually depend on i_1 only if i_2 reads the *same* memory location written by i_1. However, unless both memory operands use absolute addresses, it is hard to determine statically if the two effective addresses point to the same memory location. In our future work, we plan to use simple data flow analysis to relax this condition. Besides instructions with memory operands, this condition should also be checked for instructions with implicitly accessed memory locations, e.g., push and pop. The conditions for WAR and WAW dependencies are analogous. If no conflict is found between two instructions, then there is no constraint in their execution order.

Figure 9.2a shows the code of a basic block that contains a non-intended gadget, and Fig. 9.3 its corresponding dependence DAG. Instructions not connected via a direct edge are independent, and have no constraint in their relative execution order. Given the dependence DAG of a basic block, the possible orderings of its

---

[2] stosb (Store Byte to String) copies the least significant byte from the eax register to the memory location pointed by the edi register and increments edi's value by one. The rep prefix repeats this instruction until ecx's value reaches zero, while decreasing it after each repetition.

a   **b**

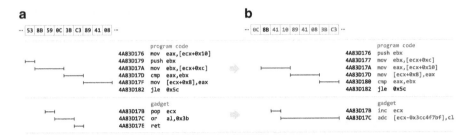

Fig. 9.2 Example of how intra basic block instruction reordering can affect a non-intended gadget

Fig. 9.3 Dependence graph
of the basic block shown in
Fig. 9.2

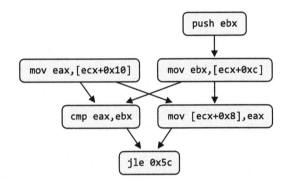

instructions correspond to the different topological sorting arrangements of the graph [69]. Figure 9.2b shows one of the possible alternative orderings of the original code. The locations of all but one of the instructions and the values of all but one of the bytes have changed, eliminating the non-intended gadget contained in the original code. Although a new gadget has appeared a few bytes further into the block, (ending again with a `ret` instruction at byte `C3`), an attacker cannot depend on it since alternative orderings will shift it to other locations, and some of its internal instructions will always change (e.g., in this example, the useful `pop ecx` is gone). In fact, the `ret` instruction can be eliminated altogether using atomic instruction substitution.

An underlying assumption we make here is that basic block boundaries will not change at runtime. If a computed control transfer instruction targets a basic block instruction other than its first, then reordering may break the semantics of the code. Although this may seem restrictive, we note that throughout our evaluation we did not encounter any such case. For compiler-generated code, IDA Pro is able to compute all jump targets even for computed jumps based on the PE relocation information. In the most conservative case, users may choose to disable instruction reordering and still benefit from the randomization of the other techniques—Sect. 9.5 includes results for each technique individually.

Fig. 9.4 Example of instruction reordering in the register preservation code at the preamble and epilogue of a function

a

```
4A834B3B   0   push ebx
4A834B3C  -4   push esi
4A834B3D  -8   mov  ebx,ecx
4A834B3F  -8   push edi
4A834B40  -C   mov  esi,edx
...
4A834B7C  -C   pop  edi
4A834B7D  -8   pop  esi
4A834B7E  -4   pop  ebx
4A834B7F   0   ret
```

b

```
4A834B3B   push edi
4A834B3C   push ebx
4A834B3D   push esi
4A834B3E   mov  ebx,ecx
4A834B40   mov  esi,edx
...
4A834B7C   pop  esi
4A834B7D   pop  ebx
4A834B7E   pop  edi
4A834B7F   ret
```

9.4.2.2 Reordering of Register Preservation Code

The calling convention followed by the majority of compilers for Windows on x86 architectures, similarly to Linux, specifies that the ebx, esi, edi, and ebp registers are callee-saved [32]. The remaining general purpose registers, known as scratch or volatile registers, are free for use by the callee without restrictions. Typically, a function that needs to use more than the available scratch registers, preserves any non-volatile registers before modifying them by storing their values on the stack. This is usually done at the function prologue through a series of push instructions, as in the example of Fig. 9.4a, which shows the very first and last instructions of a function. At the function epilogue, a corresponding series of pop instructions restores the saved values from the stack, right before returning to the caller.

Sequences that contain pop instructions followed by ret are among the most widely used gadgets found in ROP exploits, since they allow the attacker to load registers with values that are supplied as part of the injected payload [64]. The order of the pop instructions is crucial for initializing each register with the appropriate value. For example, loading 01020304 to esi and DEADC0DE to ebx using the gadget pop esi; pop ebx; ret; found in the epilogue of the function in Fig. 9.4, would require the following arrangement in the ROP payload:

```
.. |7D 6B 83 4A|04 03 02 01|DE C0 AD DE|B3 02 83 4A| ..
   | gdgt addr |    esi    |    ebx    | next gdgt |
```

As seen in the function prologue, the compiler stores the values of the callee-saved registers in arbitrary order, and sometimes the relevant push instructions are interleaved with instructions that use previously-preserved registers. At the function epilogue, the saved values are pop'ed from the stack in *reverse* order, so that they end up to the proper register. Consequently, as long as the saved values are restored in the right order, their actual order on the stack is irrelevant. Based on this observation, we can randomize the order of the push and pop instructions of register preservation code by maintaining the first-in-last-out order of the stored values, as shown in Fig. 9.4b. In this example, there are six possible orderings of the three pop instructions, which means that any assumption that the attacker may make about which registers will hold the two supplied values, will be correct with a probability of one in six (or one in three, if only one register needs to be initialized). In case only two registers are preserved, there are two possible orderings, allowing the gadget to operate correctly half of the time.

This transformation is applied conservatively, only to functions with accurately disassembled prologue and epilogue code. To make sure that we properly match the push and pop instructions that preserve a given register, we monitor the stack pointer delta throughout the whole function, as shown in the second column of Fig. 9.4a. If the deltas at the prologue and epilogue do not match, e.g., due to call sites with unknown calling conventions throughout the function, or indirect manipulation of the stack pointer, then no randomization is applied. As shown in Fig. 9.4b, any non-preservation instructions in the function prologue are reordered along with the push instructions by maintaining any interdependencies, as discussed in the previous section. For functions with multiple exit points, the preservation code at all epilogues should match the function's prologue. Note that there can be multiple push and pop pairs for the same register, in case the register is preserved only throughout some of the execution paths of a function.

9.4.3   Register Reassignment

During the register allocation phase, the compiler assigns the arbitrarily many variables of the higher-level program into the much smaller set of registers that are available in the target processor architecture. Although the program points at which a certain variable should be stored in a register or spilled into memory are chosen by sophisticated allocation algorithms, the actual name of the general purpose register that will hold a particular variable is mostly an arbitrary choice. That is, whenever a new variable needs to be mapped to a register, the compiler can pick any of the available registers at that point to hold it. As a result, the actual register assignment throughout the code of a given compiled binary is just one among many possible register assignments. Based on this observation, we can reassign the names of the register operands in the existing code according to a different—but equivalent— register assignment, without affecting the semantics of the original code. When considering each gadget as an autonomous code sequence, this transformation can alter the outcome of many gadgets, which will now read or modify different registers than those assumed by the attacker.

Due to the much higher cost of memory accesses compared to register accesses, compilers strive to map as many variables as possible to the available registers. Consequently, at any point in a large program, multiple registers are usually in use, or *live* at the same time. Given the control flow graph (CFG) of a compiled program, a register r is *live* at a program point p iff there is a path from p to a use of r that does not go through a definition of r. The *live range* of r is defined as the set of program points where r is live, and can be represented as a subgraph of the CFG [18]. Since the same register can hold different variables at different points in the program, a register can have multiple disjoint live regions in the same CFG.

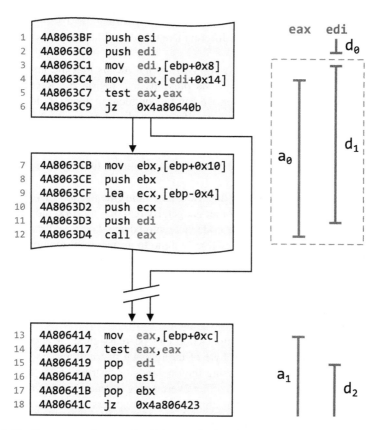

Fig. 9.5 The live ranges of eax and edi in part of a function. The two registers can be swapped in all instructions throughout their parallel, self-contained regions a_0 and d_1 (lines 3–12)

For each correctly identified function, we compute the live ranges of all registers used in its body by performing liveness analysis [12] directly on the machine code. Given the CFG of the function and the sets $use[i]$ and $def[i]$ for each instruction i, we derive the sets $in[i]$ and $out[i]$ with the registers that are *live-in* and *live-out* at each instruction. For this purpose, we use a modified version of a standard live-variable analysis algorithm [12, Fig. 9.16] that computes the *in* and *out* sets at the instruction level, instead of the basic block level. The algorithm computes the two sets by iteratively reaching a fixed point for the following data-flow equations: $in[i] = use[i] \cup (out[i] - def[i])$ and $out[i] = \bigcup\{in[s] : s \in succ[i]\}$, were $succ[i]$ is the set of all possible successors of instruction i.

Figure 9.5 shows part of the CFG of a function and the corresponding live ranges for eax and edi. Initially, we assume that all registers are live, since some of them may hold values that have been set by the caller. In this example, edi is live when entering the function, and the push instruction at line 2 stores (uses) its current value on the stack. The following mov instruction initializes (defines) edi, ending its previous live range (d_0). Note that although a live range is a sub-graph of the CFG, we illustrate and refer to the different live ranges as linear regions for the sake of convenience.

The next definition of edi is at line 15, which means that the last use of its previous value at line 11 also ends its previous live region d_1. Region d_1 is a *self-contained* region, within which we can be confident that edi holds the same variable. The eax register also has a self-contained live region (a_0) that runs in parallel with d_1. Conceptually, the two live ranges can be extended to share the same boundaries. Therefore, the two registers can be swapped across all the instructions located within the boundaries of the two regions, without altering the semantics of the code.

The call eax instruction at line 12 can be conveniently used by an attacker for calling a library function or another gadget. By reassigning eax and edi across their parallel live regions, any ROP code that would depend on eax for transferring control to the next piece of code, will now jump to an incorrect memory location, and probably crash. For code fragments with just two parallel live regions, an attacker can guess the right register half of the times. In many cases though, there are three or more general purpose registers with parallel live regions, or other available registers that are live before or after another register's live region, allowing for a higher number of possible register assignments.

The registers used in the original code can be reassigned by modifying the ModR/M and sometimes the SIB byte of the relevant instructions. As in previous code transformations, besides altering the operands of instructions in the existing code, these modifications can also affect overlapping instructions that may be part of non-intended gadgets. Note that implicitly used registers in certain instructions cannot be replaced. For example, the one-byte "move data from string to string" instruction (movs) always uses esi and edi as its source and destination operands, and there is no other one-byte instruction for achieving the same operation using a different set of registers [10]. Consequently, if such an instruction is part of the live region of one of its implicitly used registers, then this register cannot be reassigned throughout that region. For the same reason, we exclude esp from liveness analysis.

Finally, although calling conventions are followed for most of the functions, this is not always the case, as compilers are free to use any custom calling convention for private or static functions. Most of these cases are conservatively covered through a bottom-up call analysis that discovers custom register arguments and return value registers. First, all the external function definitions found in the import table of the DLL are marked as level-0 functions. IDA Pro can effectively distinguish between different calling conventions that these external functions may follow, and reports their declaration in the C language. Thus, in most cases, the register arguments and the return value register (if any) for each of the level-0 functions are known. For any call instruction to a level-0 function, its register arguments are added to call's set of implicitly read registers, and its return value registers are added to call's set of implicitly written registers.

In the next phase, level-1 functions are identified as the set of functions that call only level-0 functions or no other function. Any registers read by a level-1 function, without prior writing them, are marked as its register arguments. Similarly, any registers written and not read before a return instruction are marked as return value registers. Again, the sets of implicitly read and written register of all the `call` instructions to level-1 functions are updated accordingly. Similarly, level-2 functions are the ones that call level-1 or level-0 functions, or no other function, and so on. The same process is repeated until no more function levels can be computed. The intuition behind this approach is that private functions, which may use non-standard calling conventions, are called by other functions in the same DLL and, in most cases, not through computed call instructions.

9.5 Randomization Analysis

A crucial aspect for the effectiveness of in-place code randomization is the randomization coverage in terms of what percentage of the gadgets found in an executable can be safely randomized. A gadget may remain intact for one of the following reasons: (a) it is part of data embedded in a code segment, (b) it is part of code that could not be disassembled, or (c) it is not affected by any of our transformations. In this section, we explore the randomization coverage of our prototype implementation using a large data set of 5,235 PE files (both DLL and EXE), detailed in Table 9.1. For each PE file, we first pinpoint all gadgets contained in its executable sections. We consider as a gadget [61] any intended or unintended instruction sequence that ends with an indirect control transfer instruction, and which does not contain (a) a privileged or invalid instruction (can occur in non-intended instruction sequences), and (b) a control transfer instruction other than its final one, with the exception of indirect `call` (can be used in the middle of a gadget for calling a library function). We assume a maximum gadget length of five instructions, which is typical for existing ROP code implementations [20,61]. The larger the length of the gadget, the higher the probability that at least one of its instructions will be affected. However, for larger gadgets, it is possible that the modified part of the gadget may be irrelevant for the purpose of the attacker. For example, if only the first instruction of the gadget `inc eax; pop ebx; ret;` is randomized, this will not affect any ROP code that either does not rely on the

Table 9.1 Modifiable (eliminated vs. broken) gadgets for a collection of various PE files

Software	PE Files	Total	Modifiable %	Eliminated %	Broken %
Adobe Reader 9	43	1,250,959	75.4	8.7	66.7
Firefox 4	28	458,760	83.0	12.4	70.6
iTunes 10	75	396,478	74.0	8.0	66.0
Windows XP SP3	1,698	8,305,177	77.7	9.3	68.4
Windows 7 SP1	3,391	16,951,300	76.5	9.7	66.8
Total	5,235	27,362,674	**76.9**	**9.5**	**67.4**

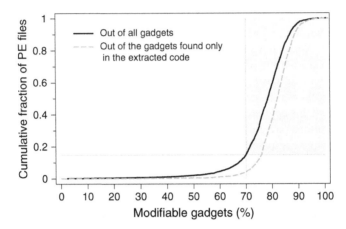

Fig. 9.6 Percentage of modifiable gadgets for a set of 5,235 PE files (detailed in Table 9.1). Indicatively, for the upper 85% of the files, more than 70% of *all* gadgets in the executable segments of each PE file can be modified (*shaded area*)

value of `eax` at that point, or uses the shorter gadget `pop ebx; ret;` directly. For this reason, we consider *all* different subsequences with length between two to five instructions as separate gadgets.

Figure 9.6 shows the percentage of modifiable gadgets out of *all* gadgets found in the executable sections of each PE file (solid line), as a cumulative fraction of all PE files in the data set. In about 85% of the PE files, more that 70% of the gadgets can be randomized by our code transformations. Many of the unmodified gadgets are located in parts of code that have not been extracted by IDA Pro, and which consequently will never be affected by our transformations. When considering only the gadgets that are contained within the disassembled code regions on which code randomization can be applied, the percentage of affected gadgets slightly increases (dashed line). Given that we do not take into account code blocks that have been identified by IDA Pro using speculative methods, this shows that the use of a more sophisticated code extraction mechanism will increase the number of gadgets that can be modified. Figure 9.7 shows the total percentage of gadgets modified by each code transformation technique for the same data set. Note that a gadget can be modified by more than one technique. Overall, the total percentage of modifiable gadgets across all PE files is about 76.9%, as shown in Table 9.1.

We identify two qualitatively different ways in which a code transformation can impact a gadget. As discussed in Sect. 9.4.1, a gadget can be *eliminated*, if any of the applied transformations removes completely its final control transfer instruction. If the final control transfer instruction remains intact, a gadget can then be *broken*, if at least one of its internal instructions is altered, and the CPU and memory state after its execution is different than the original, i.e., the outcome of its computation is not the same. As shown in Table 9.1, in the average case, about 9.5% of *all* gadgets contained in a PE file can be rendered completely unusable. For a vulnerable

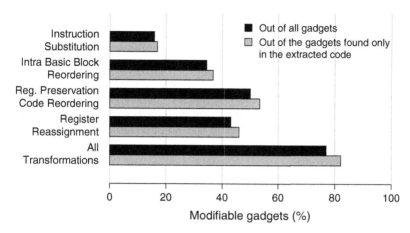

Fig. 9.7 Percentage of modifiable gadgets according to the different code transformations

application, this already removes about one in ten of the available gadgets for the construction of ROP code. Although the rest of the modifiable gadgets (67.4%) is not eliminated, they can be "broken" by probabilistically modifying one or more of their instructions.

9.6   Correctness and Performance

One of the basic principles of our approach is that the different in-place code randomization techniques should be applied cautiously, without breaking the semantics of the program. A straightforward way to verify the correctness of our code transformations is to apply them on existing code and compare the outcome before and after modification. Simply running a randomized version of a third-party application and verifying that it behaves in the expected way can provide a first indication. However, using this approach, it is hard to exercise a significant part of the code, and potentially incorrect modifications may go unnoticed.

For this purpose, we used the test suite of Wine [9], a compatibility layer that allows Windows applications to run on Unix-like operating systems. Wine provides alternative implementations of the DLLs that comprise the Windows API, and comes with an extensive test suite that covers the implementations of most of the functions exported by the core Windows DLLs. Each function is executed multiple times using various inputs that test different conditions, and the outcome of each execution is compared against a known, expected result. We ported the test code for about one third of the 109 DLLs included in the test suite of Wine v1.2.2, and used it directly on the actual DLLs gathered from a Windows 7 installation. Using multiple randomized versions of each tested DLL, we verified that in all runs, all tests completed successfully.

We took advantage of the extensive and diverse code execution coverage of this experiment to also evaluate the impact of in-place code randomization to the runtime performance of the modified code. Among the different code transformations, instruction reordering is the only one that could potentially introduce some non-negligible overhead, given that sometimes the chosen ordering may be sub-optimal. We measured the overall CPU user time for the completion of all tests by taking the average time across multiple runs, using both the original and the randomized versions of the DLLs. In all cases, there was no observable difference in the two times, within measurement error.

9.7 Effectiveness Against Real-World ROP Exploits

9.7.1 ROP Exploits and Generic ROP Payloads

We evaluated the effectiveness of in-place code randomization using publicly available ROP exploits against vulnerable Windows applications [1, 3, 4], as well as generic ROP payloads based on commonly used DLLs [8, 24]. These seven different ROP code implementations, listed in Table 9.2, bypass Windows DEP and execute a second-stage shellcode, as described in Sect. 9.2, and work even in the latest version of Windows, with DEP and ASLR enabled. We first verified that all exploits and payloads succeed by testing them against installations of the vulnerable applications on a Windows 7 SP1 virtual machine, using a shellcode that just spawns calc.exe. The ROP code used in the three exploits is implemented with gadgets from one or a few DLLs that do not support ASLR, as shown in the second column of Table 9.2. The number of unique gadgets used in each case varies between 10–18, and typically a large part of the gadgets is repeatedly executed at many points throughout the ROP code. When replacing the original non-ASLR DLLs of each application with randomized versions, in all cases the exploits were rendered unsuccessful. Similarly, we used a custom application to test the generic ROP payloads and verified that the ROP code did not succeed when the corresponding DLL was randomized.

Table 9.2 ROP exploits [1,3,4] and generic ROP payloads [8,24] tested on Windows 7 SP1

ROP exploit/payload	Gadgets in non-ASLR DLLS	Modifiable (total %: Broken % Eliminated %)			Unique Gadgets Used: Modifiable (Broken, Elim.)		Combina-tions
Adobe Reader [1]	36,760	28,637 (77.9:	70.1	7.8)	11: 6	(5, 1)	287
Integard Pro [3]	5,137	4,027 (78.4:	70.5	7.9)	16: 10	(9, 1)	322,559
Mplayer Lite [4]	117,822	104,671 (88.8:	70.0	18.8)	18: 7	(6, 1)	1,128,959
msvcr71.dll [8]	10,301	7,129 (69.2:	59.6	9.6)	14: 9	(8, 1)	3,317,760
msvcr71.dll [24]	10,301	7,129 (69.2:	59.6	9.6)	16: 8	(8, 0)	1,728,000
mscorie.dll [8]	1,616	1,304 (80.6:	73.5	7.1)	10: 4	(4, 0)	25,200
mfc71u.dll [24]	86,803	64,053 (73.8:	68.7	5.1)	11: 6	(6, 0)	170,496

The ROP code of the exploit against Acrobat Reader uses just 11 unique gadgets, all coming from a single non-ASLR DLL (icucnv36.dll). From these gadgets, in-place code randomization can alter six of them: one gadget is completely eliminated, while the other five broken gadgets have 2, 2, 3, 4, and 6 possible states, respectively, resulting to a total of 287 randomized states (*in addition* to the always eliminated gadget, which also alone breaks the ROP code). Even if we assume that no elimination were possible, the exploit would still succeed only in one out of the 288 (0.35%) possible instances (including the original) of the given gadget set. Considering that this is a client-side exploit, in which the attacker will probably have only one or a few opportunities for tricking the user to open the malicious PDF file, the achieved randomization entropy is quite high—always assuming that none of the gadgets could have been eliminated. As shown in Table 9.2, the number of possible randomized states in the rest of the cases is several orders of magnitude higher. This is mostly due to the larger number of broken gadgets, as well as due to a few broken gadgets with tens of possible modified states, which both increase the number of states exponentially.

Next, we explored whether the affected gadgets could be directly replaced with unmodifiable gadgets in order to reliably circumvent our technique. Out of the six affected gadgets in the Adobe Reader exploit, only four can be directly replaced, meaning that the exploit cannot be trivially modified to bypass randomization. Furthermore, two of the gadgets have only one replacement each, and both replacements are found in code regions that are not discovered by IDA Pro—both could be randomized using a more precise code extraction method. For the rest of the ROP payloads, there are at least three irreplaceable gadgets in each case.

We should note that the relatively small number of gadgets used in most of these ROP payloads is a worst-case scenario for our technique, which however not only is able to prevent these exploits, but also does not allow the attacker to directly replace all the affected gadgets. Indeed, besides the more complex ROP payloads used in the Integard and Mplayer exploits, the rest of the payloads use API functions that are already imported by a non-ASLR DLL, and simply call them directly using hard-coded addresses. This type of API invocation is much simpler and requires fewer gadgets [59] compared to ROP code like the one used in the Integard and Mplayer exploits (16 and 18 unique gadgets, respectively), which first dynamically locates a pointer to kernel32.dll (always ASLR-enabled in Windows 7) and then gets a handle to VirtualProtect.

9.7.2 Hindering Automated ROP Payload Generation

The fact that some of the randomized gadgets are not directly replaceable does not necessarily mean that the same outcome cannot be achieved using solely unmodifiable gadgets. For example, a gadget that performs an arithmetic operation and then copies the result to a memory location could be trivially replaced with two gadgets: one that does the arithmetic operation and one that copies the result.

To assess whether an attacker could construct a ROP payload resistant to in-place code randomization based on gadgets that cannot be randomized, we used Q [59] and Mona [25], two automated ROP code construction tools.

Q is a general-purpose ROP compiler that uses semantic program verification techniques to identify the functionality of gadgets, and provides a custom language, named QooL, for writing input programs. Its current implementation only supports simple QooL programs that call a single function or system call, while passing a single custom argument. In case the function to be called belongs to an ASLR-enabled DLL, Q can compute a handle to it through the import table of a non-ASLR DLL [34], when applicable. We should note that although Q currently compiles only basic QooL programs that call a single API function, this does not limit our evaluation, but on the contrary, stresses even more our technique. The simpler the programs, the fewer the gadgets used, which makes it easier for Q to generate ROP code even when our technique limits the number of available gadgets.

Mona is a plug-in for Immunity Debugger [2] that automates the process of building Windows ROP payloads for bypassing DEP. Given a set of non-ASLR DLLs, Mona searches for available gadgets, categorizes them according to their functionality, and then attempts to automatically generate four alternative ROP payloads for giving execute permission to the embedded shellcode and then invoking it, based on the `VirtualProtect`, `VirtualAlloc`, `NtSetInformationProcess`, and `SetProcessDEPPolicy` API functions (the latter two are not supported in Windows 7).

Considering the functionality of the ROP payloads generated by the two tools, Mona generates slightly more complex payloads, but its gadget composition engine is less sophisticated compared to Q's. Q generates payloads that compute a function address, construct its single argument, and call it. Payloads generated by Mona also call a single memory allocation API function (which though requires the construction of several arguments), copy the shellcode to the newly allocated area, and transfer control to it. Note that the complexity of the ROP code used in the tested exploits is even higher, since they rely on up to four different API functions [1], or "walk up" the stack to discover pointers to non-imported functions from ASLR-enabled DLLs [3, 4].

Table 9.3 shows the results of running Q and Mona on the same set of applications and DLLs used in the previous section (for applications, all non-ASLR DLLs are analyzed collectively), for two different cases: when all gadgets are available to the ROP compiler, and when only the non-randomized gadgets are available. The second case aims to build a payload that will be functional even when code randomization is applied. Although both Q and Mona were able to create payloads when applied on the original DLLs in almost all cases, they failed to construct any payload using only non-randomized gadgets in *all* cases. Although our technique was able to prevent two different tools from automatically constructing reliable ROP code, this favorable outcome does not exclude the possibility that a functional payload could still be constructed based solely on non-randomized gadgets, e.g., in a manual way or using an even more sophisticated ROP compiler.

Table 9.3 Results of running Q [59] and Mona [25] on the original non-ASLR DLLs listed in Table 9.2, and the unmodified parts of their randomized versions. In all cases, both tools failed to generate a ROP payload using solely non-randomized gadgets

	Q success		Mona success	
Application/DLL	Orig.	Rand.	Orig.	Rand.
Adobe Reader	✔	✘	✔ (VA)	✘
Integard Pro	✔	✘	✘	✘
Mplayer	✔	✘	✔ (VA)	✘
msvcr71.dll	✔	✘	✘	✘
mscorie.dll	✘	✘	✘	✘
mfc71u.dll	✔	✘	✔ (VA,VP)	✘

However, it clearly demonstrates that in-place code randomization significantly raises the bar for attackers, and makes the construction of reliable ROP code much harder, even in an automated way.

9.8 Discussion

Randomization Coverage. In-place code randomization may not always randomize a significant part of the executable address space, and it is hard to give a definitive answer on whether the remaining unmodifiable gadgets, or even some of the partially affected gadgets, would be sufficient for constructing useful ROP code. This depends on the code in the non-ASLR address space of the particular vulnerable process, as well as on the actual operations that need to be achieved using ROP code. Note that Turing-completeness is irrelevant for practical exploitation [59], and none of the gadget sets used in the tested ROP payloads is Turing-complete. For this reason, we emphasize that in-place code randomization should be used as a mitigation technique, in the same fashion as application armoring tools like EMET [48], and not as a complete prevention solution.

As previous studies [29, 59, 61] have shown, though, the feasibility of building a ROP payload is proportional to the size of the non-ASLR code base, and reversely proportional to the complexity of the desired functionality. Our experimental evaluation shows that in all cases, the space of the remaining useful gadgets after randomization is sufficiently small to prevent the automated generation of ROP payloads. At the same time, the tested ROP payloads are far from the complexity of a fully blown ROP-based implementation of the operations required for carrying out an attack, such as dumping a malicious executable on disk and executing it. Currently, this functionality is handled by the embedded shellcode, which in essence allows us to view these ROP payloads as sophisticated versions of return-to-libc. More complex ROP code will probably require a larger number of unique gadgets, some of them containing instructions that are currently not necessary, e.g., for directly invoking system calls. Given that even a singe broken gadget is enough to render ROP code ineffective, this would increase the potential of in-place code randomization.

In any case, in-place code randomization raises the bar for the attacker, and significantly complicates the construction of robust ROP code. We should stress that the randomization coverage of our prototype implementation is a lower bound for what would be possible using a more sophisticated code extraction method [51, 65]. In our future work, we also plan to relax some of the conservative assumptions that we have made in instruction reordering and register reassignment, using data flow analysis based on constant propagation.

Combining In-Place Code Randomization with Existing Techniques. Given its practically zero overhead and direct applicability on third-party executables, in-place code randomization can be readily combined with existing techniques to improve diversity and reduce overheads. For instance, compiler-level techniques against ROP attacks [47,54] increase significantly the size of the generated code, and also affect the runtime overhead. Incorporating code randomization for eliminating some of the gadgets could offer savings in code expansion and runtime overheads. Our technique is also applicable in conjunction with randomization methods based on code block reordering [16, 33, 43], to further increase randomization entropy.

In contrast to the above techniques, which modify the structure of the code image of a program by rearranging blocks of code, instruction set randomization (ISR) [13, 42] alters the instruction set that is "understood" by the underlying system. Legitimate programs are translated to a randomly chosen instruction set, and run normally on top of a randomized execution environment that supports the chosen instruction set. Any foreign code injected within a running process as a result of an attack would fail to execute correctly, because the actual instruction set used is unknown to any external observer.

Although ISR can be applied for protecting against any form of code injection, it is not effective against attacks such as return-to-libc and return-oriented programming, which are based on the reuse of code that already exists in the address space of a vulnerable process—irrespectively of the underlying instruction set. Conversely, in-place code randomization does not offer any protection against code injection attacks. Instruction set randomization breaks any assumptions about the instruction set used by a running process, while in-place code randomization breaks any assumptions about the location (and potentially the outcome, in case of non-intended code fragments) of certain instruction sequences that already exist in the address space of a process. The two techniques are thus complementary, and can be used in tandem to protect against both code injection and ROP attacks.

In-place code randomization at the binary level is not applicable for software that performs self-checksumming or other runtime code integrity checks. Although not encountered in the tested applications, some third-party programs may use such checks for hindering reverse engineering. Similarly, packed executables cannot be modified directly. However, in most third-party applications, only the setup executable used for software distribution is packed, and after installation all extracted PE files are available for randomization.

9.9 Conclusion

The increasing number of exploits against Windows applications that rely on return-oriented programming to bypass exploit mitigations such as DEP and ASLR, necessitates the deployment of additional protection mechanisms that can harden imminently vulnerable third-party applications against these threats. Towards this goal, we have presented in-place code randomization, a technique that offers probabilistic protection against ROP attacks, by randomizing the code of third-party applications using various narrow-scope code transformations.

Our approach is practical: it can be applied directly on third-party executables without relying on debugging information, and does not introduce any runtime overhead. At the same time, it is effective: our experimental evaluation using in-the-wild ROP exploits and two automated ROP code construction toolkits shows that in-place code randomization can thwart ROP attacks against widely used applications, including Adobe Reader on Windows 7, and can prevent the automated generation of ROP code resistant to randomization. Our prototype implementation is publicly available, and as part of our future work, we plan to improve its randomization coverage using more advanced data flow analysis methods, and extend it to support ELF and 64-bit executables.

9.10 Availability

Our prototype implementation is publicly available at http://nsl.cs.columbia.edu/projects/orp

Acknowledgements We are grateful to the authors of Q for making it available to us, and especially to Edward Schwartz for his assistance. We also thank Úlfar Erlingsson and Periklis Akritidis for their valuable feedback. This work was supported by DARPA and the US Air Force through Contracts DARPA-FA8750-10-2-0253 and AFRL-FA8650-10-C-7024, respectively, and by the FP7-PEOPLE-2009-IOF project MALCODE, funded by the European Commission under Grant Agreement No. 254116. Any opinions, findings, conclusions, or recommendations expressed herein are those of the authors, and do not necessarily reflect those of the US Government, DARPA, or the Air Force.

References

1. Adobe CoolType SING Table "uniqueName" Stack Buffer Overflow. http://www.exploit-db.com/exploits/16619/.
2. Immunity Debugger. http://www.immunityinc.com/products-immdbg.shtml.
3. Integard Pro 2.2.0.9026 (Win7 ROP-Code Metasploit Module). http://www.exploit-db.com/exploits/15016/.
4. MPlayer (r33064 Lite) Buffer Overflow + ROP exploit. http://www.exploit-db.com/exploits/17124/.

5. /ORDER (put functions in order). http://msdn.microsoft.com/en-us/library/00kh39zz.aspx.
6. Profile-guided optimizations. http://msdn.microsoft.com/en-us/library/e7k32f4k.aspx.
7. Syzygy - profile guided, post-link executable reordering. http://code.google.com/p/sawbuck/wiki/SyzygyDesign.
8. White Phosphorus Exploit Pack. http://www.whitephosphorus.org/.
9. Wine. http://www.winehq.org.
10. *Intel 64 and IA-32 Architectures Software Developer's Manual.* Volume 2 (2A & 2B): Instruction Set Reference, A-Z. 2011. http://www.intel.com/Assets/PDF/manual/325383.pdf.
11. M. Abadi, M. Budiu, U. Erlingsson, and J. Ligatti. Control-flow integrity. In *Proceedings of the 12th ACM conference on Computer and Communications Security (CCS)*, 2005.
12. A. V. Aho, M. S. Lam, R. Sethi, and J. D. Ullman. *Compilers: Principles, Techniques, and Tools (2nd Edition).* Addison-Wesley Longman Publishing Co., Inc., Boston, MA, USA, 2006.
13. E. G. Barrantes, D. H. Ackley, T. S. Palmer, D. Stefanovic, and D. D. Zovi. Randomized instruction set emulation to disrupt binary code injection attacks. In *Proceedings of the 10th ACM conference on Computer and Communications Security (CCS)*, 2003.
14. K. Baumgartner. The ROP pack. In *Proceedings of the 20th Virus Bulletin International Conference (VB)*, 2010.
15. E. Bhatkar, D. C. Duvarney, and R. Sekar. Address obfuscation: an efficient approach to combat a broad range of memory error exploits. In *In Proceedings of the 12th USENIX Security Symposium*, 2003.
16. S. Bhatkar, R. Sekar, and D. C. DuVarney. Efficient techniques for comprehensive protection from memory error exploits. In *Proceedings of the 14th USENIX Security Symposium*, August 2005.
17. T. Bletsch, X. Jiang, V. Freeh, and Z. Liang. Jump-oriented programming: A new class of code-reuse attack. In *Proceedings of the 6th Symposium on Information, Computer and Communications Security (ASIACCS)*, 2011.
18. F. Bouchez. *A Study of Spilling and Coalescing in Register Allocation as Two Separate Phases.* PhD thesis, École normale supérieure de Lyon, April 2009.
19. E. Buchanan, R. Roemer, H. Shacham, and S. Savage. When good instructions go bad: generalizing return-oriented programming to RISC. In *Proceedings of the 15th ACM conference on Computer and Communications Security (CCS)*, 2008.
20. S. Checkoway, L. Davi, A. Dmitrienko, A.-R. Sadeghi, H. Shacham, and M. Winandy. Return-oriented programming without returns. In *Proceedings of the 17th ACM conference on Computer and Communications Security (CCS)*, 2010.
21. S. Checkoway, A. J. Feldman, B. Kantor, J. A. Halderman, E. W. Felten, and H. Shacham. Can DREs provide long-lasting security? the case of return-oriented programming and the AVC advantage. In *Proceedings of the 2009 conference on Electronic Voting Technology/Workshop on Trustworthy Elections (EVT/WOTE)*, 2009.
22. P. Chen, H. Xiao, X. Shen, X. Yin, B. Mao, and L. Xie. DROP: Detecting return-oriented programming malicious code. In *Proceedings of the 5th International Conference on Information Systems Security (ICISS)*, 2009.
23. F. B. Cohen. Operating system protection through program evolution. *Computers and Security*, 12:565–584, Oct. 1993.
24. Corelan Team. Corelan ROPdb. https://www.corelan.be/index.php/security/corelan-ropdb/.
25. Corelan Team. Mona. http://redmine.corelan.be/projects/mona.
26. L. Davi, A.-R. Sadeghi, and M. Winandy. Dynamic integrity measurement and attestation: towards defense against return-oriented programming attacks. In *Proceedings of the 2009 ACM workshop on Scalable Trusted Computing (STC)*, 2009.
27. L. Davi, A.-R. Sadeghi, and M. Winandy. ROPdefender: A practical protection tool to protect against return-oriented programming. In *Proceedings of the 6th Symposium on Information, Computer and Communications Security (ASIACCS)*, 2011.
28. S. Designer. Getting around non-executable stack (and fix). http://seclists.org/bugtraq/1997/Aug/63.

29. T. Dullien, T. Kornau, and R.-P. Weinmann. A framework for automated architecture-independent gadget search. In *Proceedings of the 4th USENIX Workshop on Offensive Technologies (WOOT)*, 2010.

30. R. El-Khalil and A. D. Keromytis. Hydan: Hiding information in program binaries. In *Proceedings of the International Conference on Information and Communications Security, (ICICS)*, 2004.

31. Ú. Erlingsson. Low-level software security: Attack and defenses. Technical Report MSR-TR-07-153, Microsoft Research, 2007. http://research.microsoft.com/pubs/64363/tr-2007-153.pdf.

32. A. Fog. Calling conventions for different C++ compilers and operating systems. http://agner.org/optimize/calling_conventions.pdf.

33. S. Forrest, A. Somayaji, and D. Ackley. Building diverse computer systems. In *Proceedings of the 6th Workshop on Hot Topics in Operating Systems (HotOS-VI)*, 1997.

34. G. Fresi Roglia, L. Martignoni, R. Paleari, and D. Bruschi. Surgically returning to randomized lib(c). In *Proceedings of the 25th Annual Computer Security Applications Conference (ACSAC)*, 2009.

35. I. Guilfanov. Jump tables. http://www.hexblog.com/?p=68.

36. I. Guilfanov. Decompilers and beyond. Black Hat USA, 2008.

37. L. C. Harris and B. P. Miller. Practical analysis of stripped binary code. *SIGARCH Comput. Archit. News*, 33:63–68, December 2005.

38. Hex-Rays. IDA Pro Disassembler. http://www.hex-rays.com/idapro/.

39. X. Hu, T.-c. Chiueh, and K. G. Shin. Large-scale malware indexing using function-call graphs. In *Proceedings of the 16th ACM conference on Computer and Communications Security (CCS)*, 2009.

40. R. Hund, T. Holz, and F. C. Freiling. Return-oriented rootkits: bypassing kernel code integrity protection mechanisms. In *Proceedings of the 18th USENIX Security Symposium*, 2009.

41. R. Johnson. A castle made of sand: Adobe Reader X sandbox. CanSecWest, 2011.

42. G. S. Kc, A. D. Keromytis, and V. Prevelakis. Countering code-injection attacks with instruction-set randomization. In *Proceedings of the 10th ACM conference on Computer and Communications Security (CCS)*, 2003.

43. C. Kil, J. Jun, C. Bookholt, J. Xu, and P. Ning. Address space layout permutation (ASLP): Towards fine-grained randomization of commodity software. In *Proceedings of the 22nd Annual Computer Security Applications Conference (ACSAC)*, 2006.

44. S. Krahmer. x86-64 buffer overflow exploits and the borrowed code chunks exploitation technique. http://www.suse.de/~krahmer/no-nx.pdf.

45. C. Kruegel, W. Robertson, F. Valeur, and G. Vigna. Static disassembly of obfuscated binaries. In *Proceedings of the 13th USENIX Security Symposium*, 2004.

46. H. Li. Understanding and exploiting Flash ActionScript vulnerabilities. CanSecWest, 2011.

47. J. Li, Z. Wang, X. Jiang, M. Grace, and S. Bahram. Defeating return-oriented rootkits with "return-less" kernels. In *Proceedings of the 5th European conference on Computer Systems (EuroSys)*, 2010.

48. Microsoft. Enhanced Mitigation Experience Toolkit v2.1. http://www.microsoft.com/download/en/details.aspx?id=1677.

49. M. Miller, T. Burrell, and M. Howard. Mitigating software vulnerabilities, July 2011. http://www.microsoft.com/download/en/details.aspx?displaylang=en&id=26788.

50. S. S. Muchnick. *Advanced compiler design and implementation*. Morgan Kaufmann Publishers Inc., San Francisco, CA, USA, 1997.

51. S. Nanda, W. Li, L.-C. Lam, and T.-c. Chiueh. Bird: Binary interpretation using runtime disassembly. In *Proceedings of the International Symposium on Code Generation and Optimization (CGO)*, 2006.

52. Nergal. The advanced return-into-lib(c) exploits: PaX case study. *Phrack*, 11(58), Dec. 2001.

53. T. Newsham. Non-exec stack, 2000. http://seclists.org/bugtraq/2000/May/90.

54. K. Onarlioglu, L. Bilge, A. Lanzi, D. Balzarotti, and E. Kirda. G-Free: defeating return-oriented programming through gadget-less binaries. In *Proceedings of the 26th Annual Computer Security Applications Conference (ACSAC)*, 2010.

55. V. Pappas, M. Polychronakis, and A. D. Keromytis. Smashing the gadgets: Hindering return-oriented programming using in-place code randomization. In *Proceedings of the 33rd IEEE Symposium on Security & Privacy (S&P)*, May 2012.

56. M. Parkour. An overview of exploit packs (update 9) April 5 2011. http://contagiodump.blogspot.com/2010/06/overview-of-exploit-packs-update.html.

57. M. Pietrek. An in-depth look into the Win32 portable executable file format, part 2. http://msdn.microsoft.com/en-us/magazine/cc301808.aspx.

58. P. Saxena, R. Sekar, and V. Puranik. Efficient fine-grained binary instrumentation with applications to taint-tracking. In *Proceedings of the 6th annual IEEE/ACM international symposium on Code Generation and Optimization (CGO)*, 2008.

59. E. J. Schwartz, T. Avgerinos, and D. Brumley. Q: Exploit hardening made easy. In *Proceedings of the 20th USENIX Security Symposium*, 2011.

60. F. J. Serna. CVE-2012-0769: the case of the perfect info leak, Apr. 2012. http://zhodiac.hispahack.com/my-stuff/security/Flash_ASLR_bypass.pdf.

61. H. Shacham. The geometry of innocent flesh on the bone: return-into-libc without function calls (on the x86). In *Proceedings of the 14th ACM conference on Computer and Communications Security (CCS)*, 2007.

62. H. Shacham, M. Page, B. Pfaff, E.-J. Goh, N. Modadugu, and D. Boneh. On the effectiveness of address-space randomization. In *Proceedings of the 11th ACM conference on Computer and Communications Security (CCS)*, 2004.

63. Skape. Locreate: An anagram for relocate. *Uninformed*, 6, 2007.

64. Skape and Skywing. Bypassing Windows hardware-enforced DEP. *Uninformed*, 2, Sept. 2005.

65. M. Smithson, K. Anand, A. Kotha, K. Elwazeer, N. Giles, and R. Barua. Binary rewriting without relocation information. Technical report, University of Maryland, 2010. http://www.ece.umd.edu/~barua/without-relocation-technical-report10.pdf.

66. P. Solé. Defeating DEP, the Immunitiy Debugger way. http://www.immunitysec.com/downloads/DEPLIB.pdf.

67. P. Solé. Hanging on a ROPe. http://www.immunitysec.com/downloads/DEPLIB20_ekoparty.pdf.

68. P. Ször. *The Art of Computer Virus Research and Defense*. Addison-Wesley Professional, February 2005.

69. Y. L. Varol and D. Rotem. An algorithm to generate all topological sorting arrangements. *Comput. J.*, 24(1):83–84, 1981.

70. P. Vreugdenhil. Pwn2Own 2010 Windows 7 Internet Explorer 8 exploit. http://vreugdenhilresearch.nl/Pwn2Ownl2010-Windows7-InternetExplorer8.pdf.

71. D. A. D. Zovi. Mac OS X return-oriented exploitation. RECON, 2010.

72. D. A. D. Zovi. Practical return-oriented programming. SOURCE Boston, 2010.

Author Index

S. Jajodia et al. (eds.), *Moving Target Defense II: Application of Game Theory and Adversarial Modeling*, Advances in Information Security 100, DOI 10.1007/978-1-4614-5416-8, © Springer Science+Business Media New York 2013